These safety symbols are used in laboratory and field investigations in this
ing of each symbol and refer to this page often. *Remember to wash your h*

PROTECTIVE EQUIPMENT Do not begin any lab without the pr

| GOGGLES — Proper eye protection must be worn when performing or observing science activities that involve items or conditions as listed below. | APRON — Wear an approved apron when using substances that could stain, wet, or destroy cloth. | SOAP — Soap and water before removing goggles and after all lab activities. | — working with biological materials, chemicals, animals, or materials that can stain or irritate hands. |

LABORATORY HAZARDS

Symbols	Potential Hazards	Precaution	Response
DISPOSAL	contamination of classroom or environment due to improper disposal of materials such as chemicals and live specimens	• DO NOT dispose of hazardous materials in the sink or trash can. • Dispose of wastes as directed by your teacher.	• If hazardous materials are disposed of improperly, notify your teacher immediately.
EXTREME TEMPERATURE	skin burns due to extremely hot or cold materials such as hot glass, liquids, or metals; liquid nitrogen; dry ice	• Use proper protective equipment, such as hot mitts and/or tongs, when handling objects with extreme temperatures.	• If injury occurs, notify your teacher immediately.
SHARP OBJECTS	punctures or cuts from sharp objects such as razor blades, pins, scalpels, and broken glass	• Handle glassware carefully to avoid breakage. • Walk with sharp objects pointed downward, away from you and others.	• If broken glass or injury occurs, notify your teacher immediately.
ELECTRICAL	electric shock or skin burn due to improper grounding, short circuits, liquid spills, or exposed wires	• Check condition of wires and apparatus for fraying or uninsulated wires, and broken or cracked equipment. • Use only GFCI-protected outlets	• DO NOT attempt to fix electrical problems. Notify your teacher immediately.
CHEMICAL	skin irritation or burns, breathing difficulty, and/or poisoning due to touching, swallowing, or inhalation of chemicals such as acids, bases, bleach, metal compounds, iodine, poinsettias, pollen, ammonia, acetone, nail polish remover, heated chemicals, mothballs, and any other chemicals labeled or known to be dangerous	• Wear proper protective equipment such as goggles, apron, and gloves when using chemicals. • Ensure proper room ventilation or use a fume hood when using materials that produce fumes. • NEVER smell fumes directly. • NEVER taste or eat any material in the laboratory.	• If contact occurs, immediately flush affected area with water and notify your teacher. • If a spill occurs, leave the area immediately and notify your teacher.
FLAMMABLE	unexpected fire due to liquids or gases that ignite easily such as rubbing alcohol	• Avoid open flames, sparks, or heat when flammable liquids are present.	• If a fire occurs, leave the area immediately and notify your teacher.
OPEN FLAME	burns or fire due to open flame from matches, Bunsen burners, or burning materials	• Tie back loose hair and clothing. • Keep flame away from all materials. • Follow teacher instructions when lighting and extinguishing flames. • Use proper protection, such as hot mitts or tongs, when handling hot objects.	• If a fire occurs, leave the area immediately and notify your teacher.
ANIMAL SAFETY	injury to or from laboratory animals	• Wear proper protective equipment such as gloves, apron, and goggles when working with animals. • Wash hands after handling animals.	• If injury occurs, notify your teacher immediately.
BIOLOGICAL	infection or adverse reaction due to contact with organisms such as bacteria, fungi, and biological materials such as blood, animal or plant materials	• Wear proper protective equipment such as gloves, goggles, and apron when working with biological materials. • Avoid skin contact with an organism or any part of the organism. • Wash hands after handling organisms.	• If contact occurs, wash the affected area and notify your teacher immediately.
FUME	breathing difficulties from inhalation of fumes from substances such as ammonia, acetone, nail polish remover, heated chemicals, and mothballs	• Wear goggles, apron, and gloves. • Ensure proper room ventilation or use a fume hood when using substances that produce fumes. • NEVER smell fumes directly.	• If a spill occurs, leave area and notify your teacher immediately.
IRRITANT	irritation of skin, mucous membranes, or respiratory tract due to materials such as acids, bases, bleach, pollen, mothballs, steel wool, and potassium permanganate	• Wear goggles, apron, and gloves. • Wear a dust mask to protect against fine particles.	• If skin contact occurs, immediately flush the affected area with water and notify your teacher.
RADIOACTIVE	excessive exposure from alpha, beta, and gamma particles	• Remove gloves and wash hands with soap and water before removing remainder of protective equipment.	• If cracks or holes are found in the container, notify your teacher immediately.

Authors and Contributors

Authors

American Museum of Natural History
New York, NY

Michelle Anderson, MS
Lecturer
The Ohio State University
Columbus, OH

Juli Berwald, PhD
Science Writer
Austin, TX

John F. Bolzan, PhD
Science Writer
Columbus, OH

Rachel Clark, MS
Science Writer
Moscow, ID

Patricia Craig, MS
Science Writer
Bozeman, MT

Randall Frost, PhD
Science Writer
Pleasanton, CA

Lisa S. Gardiner, PhD
Science Writer
Denver, CO

Jennifer Gonya, PhD
The Ohio State University
Columbus, OH

Mary Ann Grobbel, MD
Science Writer
Grand Rapids, MI

Whitney Crispen Hagins, MA, MAT
Biology Teacher
Lexington High School
Lexington, MA

Carole Holmberg, BS
Planetarium Director
Calusa Nature Center and Planetarium, Inc.
Fort Myers, FL

Tina C. Hopper
Science Writer
Rockwall, TX

Jonathan D. W. Kahl, PhD
Professor of Atmospheric Science
University of Wisconsin-Milwaukee
Milwaukee, WI

Nanette Kalis
Science Writer
Athens, OH

S. Page Keeley, MEd
Maine Mathematics and Science Alliance
Augusta, ME

Cindy Klevickis, PhD
Professor of Integrated Science and Technology
James Madison University
Harrisonburg, VA

Kimberly Fekany Lee, PhD
Science Writer
La Grange, IL

Michael Manga, PhD
Professor
University of California, Berkeley
Berkeley, CA

Devi Ried Mathieu
Science Writer
Sebastopol, CA

Elizabeth A. Nagy-Shadman, PhD
Geology Professor
Pasadena City College
Pasadena, CA

William D. Rogers, DA
Professor of Biology
Ball State University
Muncie, IN

Donna L. Ross, PhD
Associate Professor
San Diego State University
San Diego, CA

Marion B. Sewer, PhD
Assistant Professor
School of Biology
Georgia Institute of Technology
Atlanta, GA

Julia Meyer Sheets, PhD
Lecturer
School of Earth Sciences
The Ohio State University
Columbus, OH

Michael J. Singer, PhD
Professor of Soil Science
Department of Land, Air and Water Resources
University of California
Davis, CA

Karen S. Sottosanti, MA
Science Writer
Pickerington, Ohio

Paul K. Strode, PhD
I.B. Biology Teacher
Fairview High School
Boulder, CO

Jan M. Vermilye, PhD
Research Geologist
Seismo-Tectonic Reservoir Monitoring (STRM)
Boulder, CO

Judith A. Yero, MA
Director
Teacher's Mind Resources
Hamilton, MT

Dinah Zike, MEd
Author, Consultant,
Inventor of Foldables
Dinah Zike Academy;
Dinah-Might Adventures, LP
San Antonio, TX

Margaret Zorn, MS
Science Writer
Yorktown, VA

Consulting Authors

Alton L. Biggs
Biggs Educational Consulting
Commerce, TX

Ralph M. Feather, Jr., PhD
Assistant Professor
Department of Educational
Studies and Secondary
Education
Bloomsburg University
Bloomsburg, PA

Douglas Fisher, PhD
Professor of Teacher Education
San Diego State University
San Diego, CA

Edward P. Ortleb
Science/Safety Consultant
St. Louis, MO

Series Consultants

Science

Solomon Bililign, PhD
Professor
Department of Physics
North Carolina Agricultural
and Technical State University
Greensboro, NC

John Choinski
Professor
Department of Biology
University of Central Arkansas
Conway, AR

Anastasia Chopelas, PhD
Research Professor
Department of Earth and
Space Sciences
UCLA
Los Angeles, CA

David T. Crowther, PhD
Professor of Science Education
University of Nevada, Reno
Reno, NV

A. John Gatz
Professor of Zoology
Ohio Wesleyan University
Delaware, OH

Sarah Gille, PhD
Professor
University of California
San Diego
La Jolla, CA

David G. Haase, PhD
Professor of Physics
North Carolina State
University
Raleigh, NC

Janet S. Herman, PhD
Professor
Department of Environmental
Sciences
University of Virginia
Charlottesville, VA

David T. Ho, PhD
Associate Professor
Department of Oceanography
University of Hawaii
Honolulu, HI

Ruth Howes, PhD
Professor of Physics
Marquette University
Milwaukee, WI

Jose Miguel Hurtado, Jr., PhD
Associate Professor
Department of Geological
Sciences
University of Texas at El Paso
El Paso, TX

Monika Kress, PhD
Assistant Professor
San Jose State University
San Jose, CA

Mark E. Lee, PhD
Associate Chair & Assistant
Professor
Department of Biology
Spelman College
Atlanta, GA

Linda Lundgren
Science writer
Lakewood, CO

Series Consultants, continued

Keith O. Mann, PhD
Ohio Wesleyan University
Delaware, OH

Charles W. McLaughlin, PhD
Adjunct Professor of Chemistry
Montana State University
Bozeman, MT

Katharina Pahnke, PhD
Research Professor
Department of Geology and Geophysics
University of Hawaii
Honolulu, HI

Jesús Pando, PhD
Associate Professor
DePaul University
Chicago, IL

Hay-Oak Park, PhD
Associate Professor
Department of Molecular Genetics
Ohio State University
Columbus, OH

David A. Rubin, PhD
Associate Professor of Physiology
School of Biological Sciences
Illinois State University
Normal, IL

Toni D. Sauncy
Assistant Professor of Physics
Department of Physics
Angelo State University
San Angelo, TX

Malathi Srivatsan, PhD
Associate Professor of Neurobiology
College of Sciences and Mathematics
Arkansas State University
Jonesboro, AR

Cheryl Wistrom, PhD
Associate Professor of Chemistry
Saint Joseph's College
Rensselaer, IN

Reading

ReLeah Cossett Lent
Author/Educational Consultant
Blue Ridge, GA

Math

Vik Hovsepian
Professor of Mathematics
Rio Hondo College
Whittier, CA

Series Reviewers

Thad Boggs
Mandarin High School
Jacksonville, FL

Catherine Butcher
Webster Junior High School
Minden, LA

Erin Darichuk
West Frederick Middle School
Frederick, MD

Joanne Hedrick Davis
Murphy High School
Murphy, NC

Anthony J. DiSipio, Jr.
Octorara Middle School
Atglen, PA

Adrienne Elder
Tulsa Public Schools
Tulsa, OK

Series Reviewers, continued

Carolyn Elliott
Iredell-Statesville Schools
Statesville, NC

Christine M. Jacobs
Ranger Middle School
Murphy, NC

Jason O. L. Johnson
Thurmont Middle School
Thurmont, MD

Felecia Joiner
Stony Point Ninth Grade Center
Round Rock, TX

Joseph L. Kowalski, MS
Lamar Academy
McAllen, TX

Brian McClain
Amos P. Godby High School
Tallahassee, FL

Von W. Mosser
Thurmont Middle School
Thurmont, MD

Ashlea Peterson
Heritage Intermediate Grade Center
Coweta, OK

Nicole Lenihan Rhoades
Walkersville Middle School
Walkersvillle, MD

Maria A. Rozenberg
Indian Ridge Middle School
Davie, FL

Barb Seymour
Westridge Middle School
Overland Park, KS

Ginger Shirley
Our Lady of Providence Junior-Senior High School
Clarksville, IN

Curtis Smith
Elmwood Middle School
Rogers, AR

Sheila Smith
Jackson Public School
Jackson, MS

Sabra Soileau
Moss Bluff Middle School
Lake Charles, LA

Tony Spoores
Switzerland County Middle School
Vevay, IN

Nancy A. Stearns
Switzerland County Middle School
Vevay, IN

Kari Vogel
Princeton Middle School
Princeton, MN

Alison Welch
Wm. D. Slider Middle School
El Paso, TX

Linda Workman
Parkway Northeast Middle School
Creve Coeur, MO

Teacher Advisory Board

The Teacher Advisory Board gave the authors, editorial staff, and design team feedback on the content and design of the Student Edition. They provided valuable input in the development of *Glencoe ⓘScience*.

Frances J. Baldridge
Department Chair
Ferguson Middle School
Beavercreek, OH

Jane E. M. Buckingham
Teacher
Crispus Attucks Medical
Magnet High School
Indianapolis, IN

Elizabeth Falls
Teacher
Blalack Middle School
Carrollton, TX

Nelson Farrier
Teacher
Hamlin Middle School
Springfield, OR

Michelle R. Foster
Department Chair
Wayland Union
Middle School
Wayland, MI

Rebecca Goodell
Teacher
Reedy Creek Middle School
Cary, NC

Mary Gromko
Science Supervisor K–12
Colorado Springs District 11
Colorado Springs, CO

Randy Mousley
Department Chair
Dean Ray Stucky
Middle School
Wichita, KS

David Rodriguez
Teacher
Swift Creek Middle School
Tallahassee, FL

Derek Shook
Teacher
Floyd Middle Magnet School
Montgomery, AL

Karen Stratton
Science Coordinator
Lexington School District One
Lexington, SC

Stephanie Wood
Science Curriculum Specialist, K–12
Granite School District
Salt Lake City, UT

Online Guide

connectED.mcgraw-hill.com

Your Digital Science Portal

   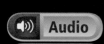         

See the science in real life through these exciting videos.

Click the link and you can listen to the text while you follow along.

Try these interactive tools to help you review the lesson concepts.

Explore concepts through hands-on and virtual labs.

These web-based challenges relate the concepts you're learning about to the latest news and research.

Digital and Print Solutions

The icons in your online student edition link you to interactive learning opportunities. Browse your online student book to find more.

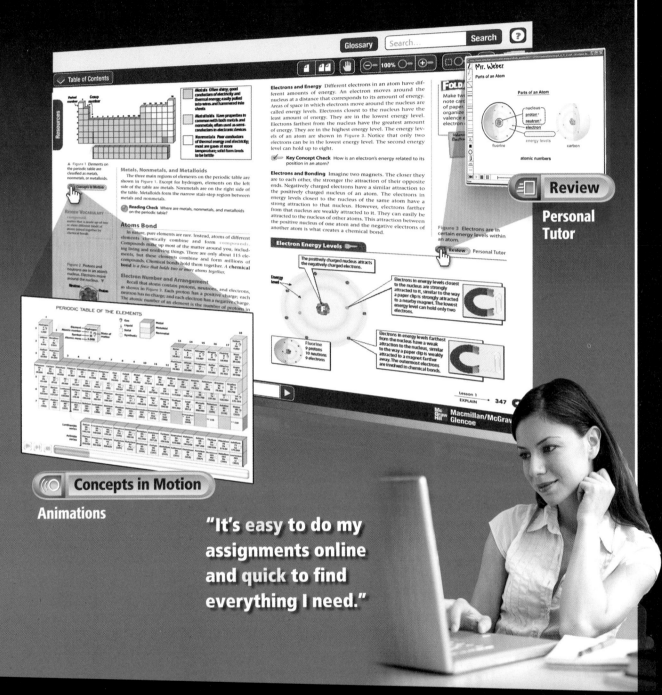

"It's easy to do my assignments online and quick to find everything I need."

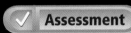

Assessment

Check how well you understand the concepts with online quizzes and practice questions.

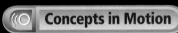

Concepts in Motion

The textbook comes alive with animated explanations of important concepts.

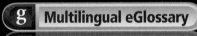

Multilingual eGlossary

Read key vocabulary in 13 languages.

ix

Treasure Hunt

START

Your science book has many features that will aid you in your learning. Some of these features are listed below. You can use the activity at the right to help you find these and other special features in the book.

- **THE BIG IDEA** can be found at the start of each chapter.
- The Reading Guide at the start of each lesson lists **Key Concepts**, vocabulary terms, and online supplements to the content.
- **Connect** icons direct you to online resources such as animations, personal tutors, math practices, and quizzes.
- **Inquiry** Labs and Skill Practices are in each chapter.
- Your **FOLDABLES** help organize your notes.

1 What four margin items can help you build your vocabulary?

2 On what page does the glossary begin? What glossary is online?

3 In which Student Resource at the back of your book can you find a listing of Laboratory Safety Symbols?

4 Suppose you want to find a list of all the Launch Labs, MiniLabs, Skill Practices, and Labs, where do you look?

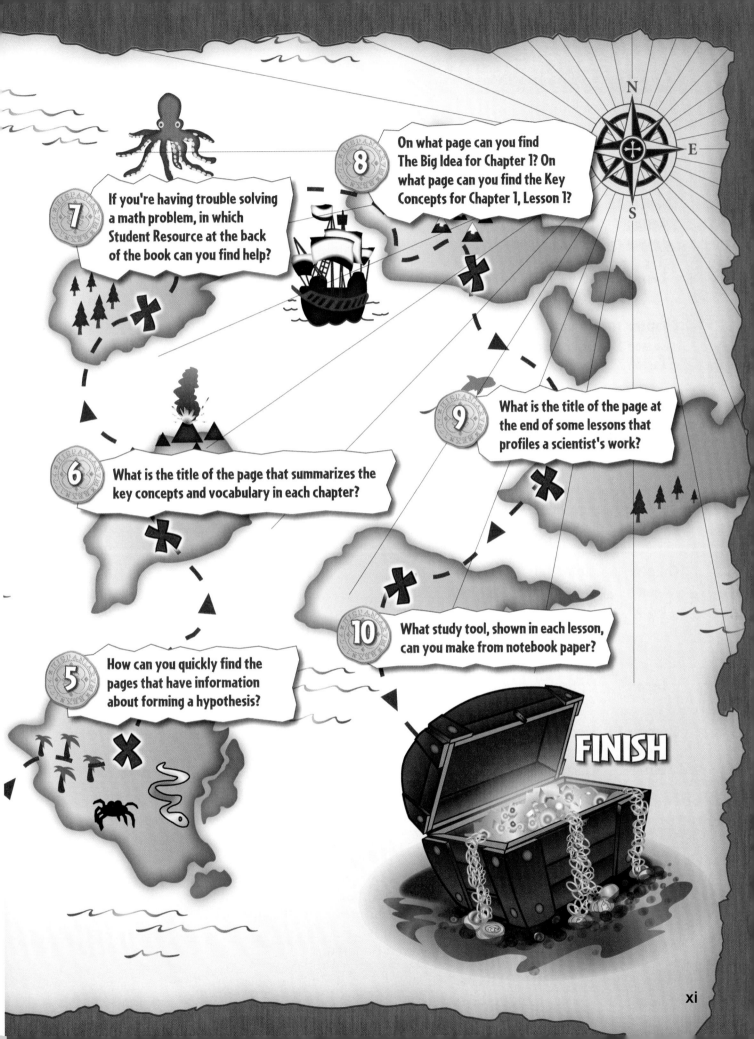

Table of Contents

Unit 5   Waves, Electricity, and Magnetism 522

Chapter 15   **Waves** 526
- Lesson 1   What are waves? 528
- Lesson 2   Wave Properties 538
  - Skill Practice   How are the properties of waves related? 545
- Lesson 3   Wave Interactions 546
  - Lab   Measuring Wave Speed 554

Chapter 16   **Sound** 562
- Lesson 1   Producing and Detecting Sound 564
- Lesson 2   Properties of Sound Waves 572
  - Skill Practice   How can you use a wind instrument to play music? ... 581
- Lesson 3   Using Sound Waves 582
  - Lab   Make Your Own Musical Instrument 590

Chapter 17   **Electromagnetic Waves** 598
- Lesson 1   Electromagnetic Radiation 600
- Lesson 2   The Electromagnetic Spectrum 608
  - Skill Practice   What's at the edge of a rainbow? 613
- Lesson 3   Using Electromagnetic Waves 614
  - Lab   Design an Exhibit for a Science Museum 624

Chapter 18   **Light** 632
- Lesson 1   Light, Matter, and Color 634
- Lesson 2   Reflection and Mirrors 642
  - Skill Practice   How can you demonstrate the law of reflection? 648
- Lesson 3   Refraction and Lenses 649
  - Skill Practice   How does a lens affect light? 659
- Lesson 4   Optical Technology 660
  - Lab   Design an Optical Illusion 668

Chapter 19   **Electricity** 676
- Lesson 1   Electric Charge and Electric Forces 678
- Lesson 2   Electric Current and Simple Circuits 689
  - Skill Practice   What effect does voltage have on a circuit? 697
- Lesson 3   Describing Circuits 698
  - Lab   Design an Elevator 706

Chapter 20   **Magnetism** 714
- Lesson 1   Magnets and Magnetic Fields 716
- Lesson 2   Making Magnets with an Electric Current 726
  - Skill Practice   How can you measure an electric current? 733
- Lesson 3   Making Electric Current with Magnets 734
  - Lab   Design a Wind-Powered Generator 742

Table of Contents

Student Resources

Science Skill Handbook **SR-2**
   Scientific Methods SR-2
   Safety Symbols SR-11
   Safety in the Science Laboratory SR-12

Math Skill Handbook **SR-14**
   Math Review SR-14
   Science Application SR-24

Foldables Handbook **SR-29**

Reference Handbook **SR-40**
   Periodic Table of the Elements SR-40

Glossary **G-2**

Index **I-2**

Credits **C-2**

Inquiry

Launch Labs

15-1	How can you make waves?	529
15-2	Which sounds have more energy?	539
15-3	What happens in wave collisions?	547
16-1	What causes sound?	565
16-2	How can sound blow out a candle?	573
16-3	Why didn't you hear the phone ringing?	583
17-1	How can you detect invisible waves?	601
17-2	How do electromagnetic waves differ?	609
17-3	How do X-rays see inside your teeth?	615
18-1	How can you make a rainbow?	635
18-2	How can you read a sign behind you?	643
18-3	What happens to light that passes from one transparent substance to another?	650
18-4	How do long-distance phone lines work?	661
19-1	How can you bend water?	679
19-2	What is the path to turn on a light?	690
19-3	How would you wire a house?	699
20-1	What does *magnetic* mean?	717
20-2	When is a wire a magnet?	727
20-3	Why does the pendulum slow down?	735

MiniLabs

15-1	How do waves travel through matter?	531
15-2	How are wavelength and frequency related?	541
15-3	How can reflection be used?	549
16-1	How do you know a sound's direction?	569
16-2	How can you hear beats?	577
16-3	How fast is sound?	585
17-1	How are electric fields and magnetic fields related?	603
17-3	How does infrared imaging work?	620
18-1	What color is that?	638
18-2	Where is the image in a plane mirror?	645
18-3	How can water move light?	652
18-4	How does a zoom lens work?	663
19-1	How can a balloon push or pull?	682
19-2	When is one more than two?	692
19-3	What else can a circuit do?	704
20-1	Where is magnetic north?	721
20-2	What is an electromagnet?	730
20-3	How many paper clips can you lift?	737

Inquiry

Inquiry Skill Practice

- **15-2** How are the properties of waves related? 545
- **16-2** How can you use a wind instrument to play music? 581
- **17-2** What's at the edge of a rainbow? 613
- **18-2** How can you demonstrate the law of reflection? 648
- **18-3** How does a lens affect light? 659
- **19-2** What effect does voltage have on a circuit? 697
- **20-2** How can you measure an electric current? 733

Inquiry Labs

- **15-3** Measuring Wave Speed 554
- **16-3** Make Your Own Musical Instrument 590
- **17-3** Design an Exhibit for a Science Museum 624
- **18-4** Design an Optical Illusion 668
- **19-3** Design an Elevator 706
- **20-3** Design a Wind-Powered Generator 742

Features

How It Works

- **16-1** Cochlear Implants 571
- **17-1** Solar Sails 607
- **19-1** Van de Graaff Generator 688

Science & Society

- **18-1** Color! 641
- **20-1** Roller Coasters 725

Careers in Science

- **15-1** Making a Computer Tsunami 537

Unit 5

WAVES, ELECTRICITY, & MAGNETISM

1660
Robert Hooke publishes the wave theory of light, comparing light's movement to that of waves in water.

1705
Francis Hauksbee experiments with a clock in a vacuum and proves that sound cannot travel without air.

1820
Danish physicist Hans Christian Ørsted publishes his discovery that an electric current passing through a wire produces a magnetic field.

1878
Thomas Edison develops a system to provide electricity to homes and businesses using locally generated and distributed direct current (DC) electricity.

1882
Thomas Edison develops and builds the first electricity-generating plant in New York City, which provides 110 V of direct current to 59 customers in lower Manhattan.

1883
The first standardized incandescent electric lighting system using overhead wires begins service in Roselle, New Jersey.

1890s
Physicist Nikola Tesla introduces alternating current (AC) by inventing the alternating current generator, allowing electricity to be transmitted at higher voltages over longer distances.

1947
Chuck Yeager becomes the first pilot to travel faster than the speed of sound.

Inquiry
Visit ConnectED for this unit's STEM activity.

Unit 5 Nature of SCIENCE

Graphs

Have you ever felt a shock from static electricity? The electric energy that you feel is similar to the electric energy you see as a flash of lightning, such as in **Figure 1**, only millions of times smaller. Scientists are still investigating what causes lightning and where it will occur. They use graphs to learn about the risk of lightning in different places and at different times. A **graph** is a type of chart that shows relationships between variables. Graphs organize and summarize data in a visual way. Three of the most common graphs are circle graphs, bar graphs, and line graphs.

Types of Graphs

Line Graphs

A line graph is used when you want to analyze how a change in one variable affects another variable. This line graph shows how the average number of lightning flashes changes over time in Illinois. Time is plotted on the x-axis. The average numbers of lightning flashes are plotted on the y-axis. Each dot, called a data point, indicates the average number of flashes recorded during that hour. A line connects the data points so a trend can be analyzed.

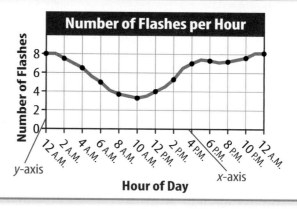

Bar Graphs

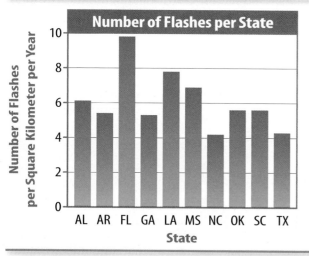

When you want to compare amounts in different categories, you use a bar graph. The horizontal axis often contains categories instead of numbers. This bar graph shows the average number of lightning flashes that occur in different states. On average, about 9.8 lightning flashes strike each square kilometer of land in Florida every year. Florida has more lightning flashes per square kilometer than all other states shown on the graph.

Circle Graphs

If you want to show how the parts of something relate to the whole, use a circle graph. This circle graph shows the average percentage of lightning flashes each U.S. region receives in a year. The graph shows that the northeastern region of the United States receives about 14 percent of all lightning flashes that strike the country each year. From the graph, you can also determine that the southeast receives the most lightning in a given year.

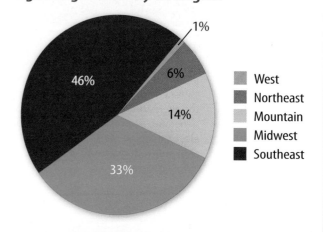

Lightning Flashes by US Region

524 • Nature of Science

Line Graphs and Trends

Suppose you are planning a picnic in an area that experiences quite a bit of lightning. When would be the safest time to go? First, you gather data about the average number of lightning flashes per hour. Next, you plot the data on a line graph and analyze trends. Trends are patterns in data that help you find relationships among the data and make predictions.

Follow the orange line on the line graph from 12 A.M. to 10 A.M. in **Figure 2**. Notice that the line slopes downward, indicated by the green arrow. A downward slope means that as measurements on the *x*-axis increase, measurements on the *y*-axis decrease. So, as time passes from 12 A.M. to 10 A.M., the number of lightning flashes decreases.

If you follow the orange line on the line graph from 12 P.M. to 5 P.M., you will notice that the line slopes upward. This is indicated by the blue arrow. An upward slope means that as the measurements on the *x*-axis increase, the measurements on the *y*-axis also increase. So, as time passes from 12 P.M. to 5 P.M., the number of lightning flashes increases.

The line graph shows you that between about 8 A.M. and 12 P.M., you would have the least risk of lightning during your picnic.

▲ **Figure 1** Scientists study lightning to get a better understanding of what causes it and to predict when it will occur.

Figure 2 The slope of a line in a line graph shows the relationship between the variables on the *x*-axis and the variables on the *y*-axis. ▼

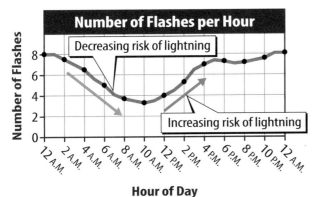

Inquiry MiniLab — 25 minutes

When does lightning strike?

Meteorologists in New Mexico collected data on the number of lightning flashes throughout the day. How can you use a line graph to plan the safest day trip?

1. Make a line graph of the data in the table.
2. Find the trends that show when the risk for lightning is increasing and when it is decreasing.

Hour	# of Flashes
12:00 A.M.	2
3:00 A.M.	2
6:00 A.M.	2
9:00 A.M.	2
12:00 P.M.	5
3:00 P.M.	21
6:00 P.M.	36
9:00 P.M.	22
12:00 A.M.	2

Analyze and Conclude

Decide How could you use your graph to plan a day trip in New Mexico with the least risk of lightning?

Chapter 15

Waves

 How do waves travel through matter?

Inquiry **What causes waves?**

Waves are actually energy moving through matter. Think about the amount of energy this wave must be carrying.

- What do you think caused this giant wave?
- Do you think this is the only large wave in the area?
- How do you think this wave moves through water?

Get Ready to Read

What do you think?

Before you read, decide if you agree or disagree with each of these statements. As you read this chapter, see if you change your mind about any of the statements.

1. Waves carry matter as they travel from one place to another.
2. Sound waves can travel where there is no matter.
3. Waves that carry more energy cause particles in a material to move a greater distance.
4. Sound waves travel fastest in gases, such as those in the air.
5. When light waves strike a mirror, they change direction.
6. Light waves travel at the same speed in all materials.

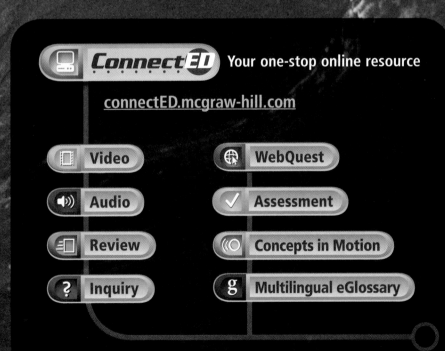

ConnectED Your one-stop online resource

connectED.mcgraw-hill.com

- Video
- WebQuest
- Audio
- Assessment
- Review
- Concepts in Motion
- Inquiry
- Multilingual eGlossary

Lesson 1

Reading Guide

Key Concepts 🔑
ESSENTIAL QUESTIONS

- What is a wave?
- How do different types of waves make particles of matter move?
- Can waves travel through empty space?

Vocabulary
wave p. 529
mechanical wave p. 531
medium p. 531
transverse wave p. 531
crest p. 531
trough p. 531
longitudinal wave p. 532
compression p. 532
rarefaction p. 532
electromagnetic wave p. 535

g Multilingual eGlossary

What are waves?

Inquiry: Why Circles?

Have you ever seen raindrops falling on a smooth pool of water? If so, you probably saw a pattern of circles forming. What are these circles? The circles are small waves that spread out from where the raindrops hit the water.

528 Chapter 15
ENGAGE

Inquiry Launch Lab

20 minutes

How can you make waves?

Oceans, lakes, and ponds aren't the only places you can find waves. Can you create waves in a cup of water?

1. Read and complete a lab safety form.
2. Add **water** to a **clear plastic cup** until it is about two-thirds full. Place the cup on a **paper towel.**
3. Explore ways of producing water waves by touching the cup. Do not move the cup.
4. Now explore ways of producing water waves without touching the cup. Do not move the cup.

Think About This

1. How did the water's surface change when you produced water waves in the cup?

2. **Key Concept** What did the different ways of producing water waves have in common?

What are waves?

Imagine a warm summer day. You are floating on a raft in the middle of a calm pool. Suddenly, a friend does a cannonball dive into the pool. You probably know what happens next—you are no longer resting peacefully on your raft. Your friend's dive causes you to start bobbing up and down on the water. You might notice that after you stop moving up and down, you haven't moved forward or backward in the pool.

Why did your friend's dive make you move up and down? Your friend created waves by jumping into the pool. *A* **wave** *is a disturbance that transfers energy from one place to another without transferring matter.* You moved up and down because these waves transferred energy.

Key Concept Check What is a wave?

A Source of Energy

The photo on the previous page shows the waves produced when raindrops fall into a pond. The impact of the raindrops on the water is the source of energy for these water waves. Waves transfer energy away from the source of the energy. **Figure 1** shows how light waves spread out in all directions away from a flame. The burning wick is the energy source for these light waves.

REVIEW VOCABULARY
energy
the ability to cause change

Figure 1 All waves, such as light waves, spread out from the energy source that produces the waves.

Lesson 1
EXPLORE
529

Figure 2 Water waves transfer energy across the pool, but not matter. As a result, the raft does not move along with the waves. Instead, the raft returns to its initial position.

The raft is at rest in its initial position.

A wave begins to lift the raft upward when it reaches the raft.

The wave transfers energy to the raft as it lifts it upward.

The wave passes the raft and continues to move across the pool.

The raft returns to its initial position after the wave passes.

Visual Check Describe what happens to the raft when the waves transfer energy to it.

Energy Transfer

Think about the waves created by a cannonball dive. When the diver hits the water, the diver's energy transfers to the water. Recall that energy is the ability to cause change. The energy transferred to the water produces waves. The waves transfer energy from the place where the diver hits the water to the place where your raft is floating. **Figure 2** shows that the energy transferred by a wave lifts your raft when the wave reaches it.

 **Reading Check** What do waves transfer from place to place?

The waves created in the pool caused you to move up and then down on your raft. As **Figure 2** shows, however, after the waves passed, you were in the same place in the pool. The waves didn't carry you along after they reached you. Waves transfer energy, but they leave matter in the same place after they pass. Because the water under your raft was not carried along with the waves, you remained in the same place in the pool.

How Waves Transfer Energy

Why wouldn't a water wave carry a raft along with it? How do waves transfer energy without transferring matter? Think about the diver hitting the water. Like all materials, water is made of tiny particles. When the diver hit the water, the impact of the diver exerted a force on water particles. The force of the impact transferred energy to the water by pushing and pulling on water particles. These particles then pushed and pulled on neighboring water particles, transferring energy outward from the point of impact. In this way, the energy of the falling diver is transferred to the water. This energy then travels through the water, from particle to particle, as a wave.

Mechanical Waves

A water wave is an example of a mechanical wave. *A wave that can travel only through matter is a* **mechanical wave.** Mechanical waves can travel through solids, liquids, and gases. They cannot travel through a vacuum. *A material in which a wave travels is called a* **medium.** Two types of mechanical waves are transverse waves and longitudinal waves.

Transverse Waves

You can make a wave on a rope by shaking the end of the rope up and down, as shown in **Figure 3.** A wave on a rope is a transverse wave. *A* **transverse wave** *is a wave in which the disturbance is perpendicular to the direction the wave travels.* **Figure 3** shows that the particles in the rope move up and down while the wave travels horizontally. The up-and-down movement of the rope particles is at right angles to the direction of wave movement.

 Key Concept Check How do particles move in a transverse wave?

The dotted line in **Figure 3** shows the position of the rope before you start shaking it. This position is called the rest position. The transverse wave on the rope has high points and low points. *The highest points on a transverse wave are* **crests.** *The lowest points on a transverse wave are* **troughs.**

Inquiry MiniLab   15 minutes

How do waves travel through matter?

When a wave travels through matter, energy transfers from particle to particle. How do particles move in different types of waves?

1. Read and complete a lab safety form.
2. Tie a piece of **yarn** around a **rope.** Stretch the rope on the floor between you and a partner. Make a transverse wave by moving the rope side to side on the floor. Observe how the yarn moves.
3. Tie a piece of yarn around one coil near the middle of a **metal spring toy.** Stretch the toy between you and a partner on the floor. Sharply push one end of the spring toy forward. Observe how the yarn moves.

Analyze and Conclude

1. **Compare and contrast** the motion of the yarn on the rope and the motion of the yarn on the spring.
2. **Key Concept** Write a statement about the motion of the particles of a medium as a wave passes.

Figure 3 In a transverse wave, particles move at right angles to the direction the wave travels.

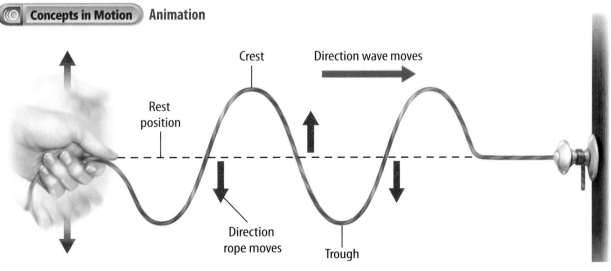

Longitudinal Waves

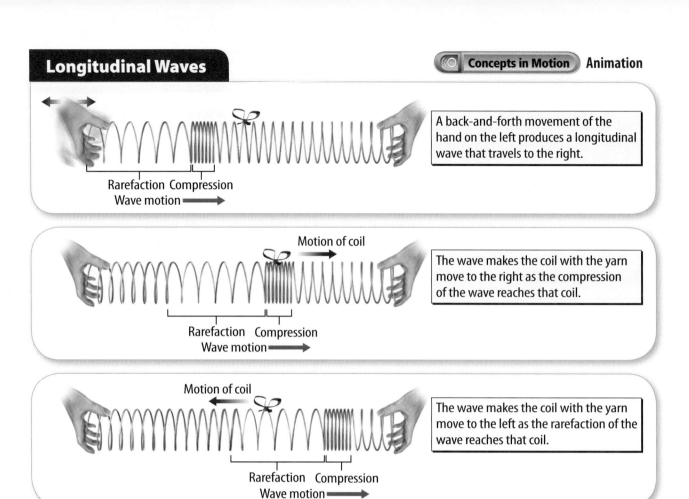

Figure 4 A longitudinal wave travels along the spring when the hand moves back and forth. The coils of the spring move back and forth along the same direction that the wave travels.

Longitudinal Waves

Another type of mechanical wave is a longitudinal (lahn juh TEWD nul) wave. *A* **longitudinal wave** *makes the particles in a medium move parallel to the direction that the wave travels.* A longitudinal wave traveling along a spring is shown in **Figure 4.** As the wave passes, the coils of the spring move closer together, then farther apart, and then return to their original positions. The coils move back and forth parallel to the direction that the wave moves.

Before a wave is produced in the spring, the coils are the same distance apart. This is the rest position of the spring. **Figure 4** shows that the wave produces regions in the spring where the coils are closer together than they are in the rest position and regions where they are farther apart. *The regions of a longitudinal wave where the particles in the medium are closest together are* **compressions.** *The regions of a longitudinal wave where the particles are farthest apart are* **rarefactions.**

Key Concept Check How do particles move in a longitudinal wave?

Vibrations and Mechanical Waves

If you hit a large drum with a drumstick, the surface of the drum vibrates, or moves up and down. A vibration is a back-and-forth or an up-and-down movement of an object. Vibrating objects, such as a drum or a guitar string, are the sources of energy that produce mechanical waves.

 Reading Check What produces mechanical waves?

One Wave per Vibration Suppose you move the end of a rope down, up, and back down to its original position. Then you make a transverse wave with a crest and a trough, as shown in the top of **Figure 5**. The up-and-down movement of your hand is one vibration. One vibration of your hand produces a transverse wave with one crest and one trough.

Similarly, a single back-and-forth movement of the end of a spring produces a longitudinal wave with one compression and one rarefaction. In both cases, the vibration of your hand is the source of energy that produces the transverse or longitudinal wave.

Vibrations Stop—Waves Go Now imagine that you move the end of the rope up and down several times. The motion of your hand transfers energy to the rope and produces several crests and troughs, as the bottom of **Figure 5** shows. As long as your hand keeps moving up and down, energy transfers to the rope and produces waves.

When your hand stops moving, waves no longer are produced. However, as shown in both parts of **Figure 5,** waves produced by the earlier movements of your hand continue to travel along the rope. This is true for any vibrating object. Waves can keep moving even after the object stops vibrating.

 Reading Check If your hand makes four vibrations, how many waves are created?

Figure 5 Vibrations produce waves that keep traveling even when the vibrations stop.

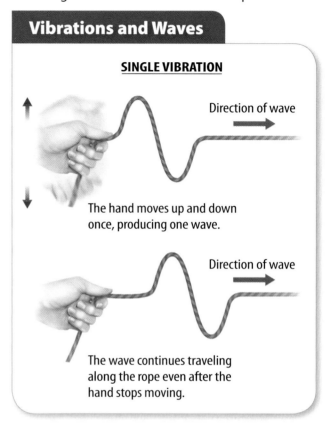

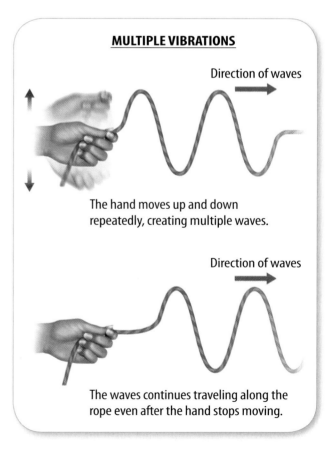

Lesson 1
EXPLAIN
533

Types of Mechanical Waves

All mechanical waves travel only in matter. Sound waves, water waves, and waves produced by earthquakes are mechanical waves that travel in different mediums. **Table 1** shows examples of these mechanical waves.

Visual Check What are the mediums for each of the mechanical waves in **Table 1**?

Table 1 Types of Mechanical Waves

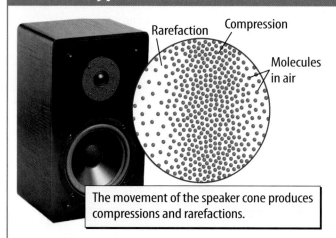

The movement of the speaker cone produces compressions and rarefactions.

Sound Waves
- Sound waves are longitudinal waves that travel in solids, liquids, and gases.
- A sound wave is made of a series of compressions and rarefactions.
- A paper cone inside the speaker vibrates in and out, producing sound waves.
- The speaker makes a compression when the speaker cone pushes air molecules together as it moves outward.
- The speaker makes a rarefaction when the speaker cone moves inward and the air molecules spread out.

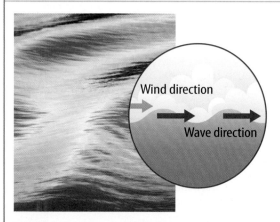

Water Waves
- Water waves are a combination of transverse waves and longitudinal waves.
- Wind produces most waves in oceans and lakes by pushing on the surface of the water.

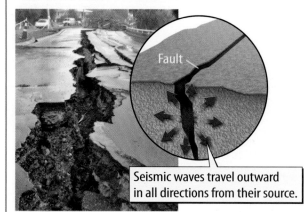

Seismic waves travel outward in all directions from their source.

Seismic Waves
- Waves in Earth's crust, called seismic (SIZE mihk) waves, cause earthquakes.
- Seismic waves are mechanical waves that travel within Earth and on Earth's surface.
- There are both longitudinal and transverse seismic waves.
- In some places, parts of Earth's upper layers can move along a crack called a fault.
- The movement of Earth's upper layers along a fault produces seismic waves.

Electromagnetic Waves

The Sun gives off light that travels through space to Earth. Like sound waves and water waves, light is also a type of wave. However, light is not a mechanical wave. A mechanical wave cannot travel through the space between the Sun and Earth. Light is an electromagnetic wave. *An* **electromagnetic wave** *is a wave that can travel through empty space and through matter.*

 Key Concept Check Identify a type of wave that can travel through a vacuum.

Types of Electromagnetic Waves

In addition to light waves, other types of electromagnetic waves include radio waves, microwaves, infrared waves, and ultraviolet waves. Cell phones use microwaves that carry sounds from one phone to another. When you stand by a fire, infrared waves striking your skin cause the warmth you feel. Ultraviolet waves from the Sun cause sunburns.

Electromagnetic Waves and Objects

Every object, including you, gives off electromagnetic waves. The type of electromagnetic waves that an object gives off depends mainly on the temperature of the object. For example, you give off mostly infrared waves. Other objects near human body temperature also give off mostly infrared waves. Hotter objects, such as a piece of glowing metal, give off visible light waves as well as infrared waves. Some animals, such as the copperhead in **Figure 6,** have specialized detectors for perceiving the infrared waves given off by their prey.

Electromagnetic Waves from the Sun

Like all waves, electromagnetic waves carry energy. Scientists often call this radiant energy. Infrared and visible light waves carry about 92 percent of the radiant energy that reaches Earth from the Sun. Ultraviolet waves carry about 7 percent of the Sun's energy.

WORD ORIGIN
electromagnetic
from Greek *elektron*, means "amber" and *magnes*, means "lodestone"

Make a two-tab book and label it as shown. Use your book to organize information about mechanical and electromagnetic waves.

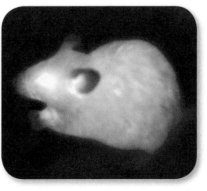

Infrared sensing pit

Figure 6 Some snakes have special organs on their heads that allow them to detect infrared waves from prey, such as mice. The right photo shows the infrared waves that a mouse gives off.

Lesson 1 Review

Visual Summary

Waves, such as those from a burning candle, the Sun, or a loudspeaker, transfer energy away from the source of the wave.

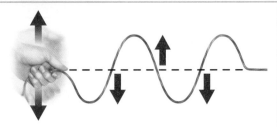

A transverse wave makes particles in a medium move perpendicular to the direction of the wave.

A longitudinal wave makes the particles in a medium move in a direction that is parallel to the direction the wave travels.

FOLDABLES

Use your lesson Foldable to review the lesson. Save your Foldable for the project at the end of the chapter.

What do you think NOW?

You first read the statements below at the beginning of the chapter.

1. Waves carry matter as they travel from one place to another.
2. Sound waves can travel where there is no matter.

Did you change your mind about whether you agree or disagree with the statements? Rewrite any false statements to make them true.

Use Vocabulary

1. **Define** *wave* in your own words.

2. **Distinguish** between a transverse wave and a longitudinal wave.

Understand Key Concepts

3. What causes a wave?
   A. a crest   C. a rope
   B. a rarefaction D. a vibration

4. **Differentiate** How are sound waves and water waves similar? How are they different?

Interpret Graphics

5. **Describe** what happens to the wave shown below when the hand stops vibrating.

Direction of waves

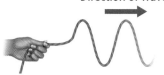

6. **Compare and Contrast** Copy and fill in the graphic organizer below to compare and contrast mechanical waves and electromagnetic waves.

Similarities	Differences

Critical Thinking

7. **Analyze** A scientist wants to analyze signals from outer space that tell her about the age of the universe. Her lab is equipped with advanced sensors that detect sound waves, radio waves, and seismic waves. Which of these sensors will provide the best information? Explain your reasoning.

8. **Predict** You are floating motionless on a rubber raft in the middle of a pool. A friend forms a wave by slapping the water every second. Will the wave carry you to the edge of the pool? Explain your answer.

Making a Computer Tsunami

CAREERS in SCIENCE

Meet Vasily Titov, a scientist working to predict the next big wave.

On the morning of December 26, 2004, a magnitude 9.3 earthquake in the Indian Ocean caused an enormous tsunami. On the other side of the world, a scientist in Seattle, Washington, sprang into action. Vasily Titov worked through the night to develop the first-ever computer model of the tsunami.

Titov is a mathematician for the National Oceanic and Atmospheric Administration (NOAA). His computer model solves equations that describe how tsunamis are produced and how they move. The model includes data about an earthquake's energy, the shape of the ocean floor, and the ocean's depth. When there is an earthquake on the ocean floor, Titov's model predicts the properties of the resulting tsunami.

A tsunami moves about as fast as a jet airplane. By the time Titov's computer model was ready, the 2004 Indian Ocean tsunami had already hit the coast and killed more than 280,000 people. But Titov hopes this new tool will help communities prepare for future tsunamis.

Now Titov is applying his computer model to the Pacific Northwest. His model shows how tsunamis could affect the coastlines of Washington, Oregon, and northern California. Emergency managers can use these results to predict when tsunami waves will reach different parts of the coast. Warnings could be issued to communities just minutes after an earthquake triggers a tsunami.

AMERICAN MUSEUM of NATURAL HISTORY

What's a tsunami?

Wind causes most ordinary water waves, but tsunamis are not ordinary waves. Underwater earthquakes usually cause these giant, destructive waves. As a tectonic plate moves upward, it raises the water above it. As this water falls back down, it forms a massive wave that spreads in all directions.

Seafloor • **Fault** • **Earthquake** • **Generation**

❶ In deep water, tsunamis usually travel faster than 250 m/s, but they are only a few centimeters high. Most ships in deep water cannot even detect a passing tsunami.

❷ As the wave approaches shore, much of its kinetic energy is transformed into gravitational potential energy. The water piles up, forming enormous walls of water that flood coastal areas.

It's Your Turn

WRITE Suppose you were a witness to the 2004 tsunami. Write a diary entry recording your experience. What was your first hint that something was wrong? What was happening around you? What did you see and hear?

Lesson 1 EXTEND

Lesson 2

Reading Guide

Key Concepts
ESSENTIAL QUESTIONS

- What are properties of waves?
- How are the frequency and the wavelength of a wave related?
- What affects wave speed?

Vocabulary
amplitude p. 539
wavelength p. 541
frequency p. 542

 Multilingual eGlossary

 Video

- BrainPOP®
- Science Video

Wave Properties

Inquiry Why So Big?

Have you ever watched a surfer on a huge wave? If so, you might have wondered why some waves are huge and why others are small. The size of a water wave depends on the energy it carries.

Inquiry Launch Lab

20 minutes

Which sounds have more energy?

Some sounds are loud and others are soft. What is the difference between loud and soft sounds?

1. Read and complete a lab safety form.
2. Place a **bowl of water** over a sheet of **white paper.**
3. Strike a **tuning fork** gently on your hand so it makes a soft sound and then quickly place its prongs in the bowl of water. Remove the tuning fork.
4. Strike the tuning fork sharply on your hand so it makes a loud sound and then quickly place its prongs in the bowl of water.

Think About This

1. **Contrast** the waves made by the tuning fork in steps 3 and 4 in your Science Journal.
2. In which step did the tuning fork transfer more energy to the water? Explain your answer.
3. **Key Concept** How are the loudness of the sounds and the vibrations of the tuning fork related?

Amplitude and Energy

Imagine you are floating on a raft in a pool and someone gently splashes the water near you, creating waves. You might barely feel these waves lift you up and down as they pass. If someone dives off the diving board, this makes waves that bounce you up and down. What's the difference between these waves?

Initially, the surface of the water you are floating on is nearly flat. This is the rest position for the water. The dive produced waves with higher crests and deeper troughs than those of the waves produced by gently splashing the water. This means that the dive caused water to move a greater distance from its rest position, producing a wave with a greater amplitude. *The **amplitude** of a wave is the maximum distance that the wave moves from its rest position.*

For any wave, the larger the amplitude, the more energy the wave carries. The wave produced by the diver hitting the water caused a greater change than the wave produced by the gentle splash. The first wave had more energy.

Reading Check What is the amplitude of a wave? How are the amplitude of a wave and energy related?

WORD ORIGIN
amplitude
from Latin *amplitudinem*, means "width"

FOLDABLES
Make a layered book from two half sheets of paper. Label it as shown. Use your book to organize your notes about the properties of waves.

- Properties of Waves
- Amplitude
- Wavelength
- Frequency

Lesson 2
EXPLORE

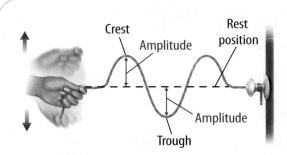
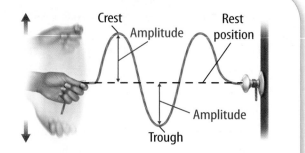

This wave has a smaller amplitude and carries less energy. This wave has a greater amplitude and carries more energy.

▲ **Figure 7** The amplitude of a transverse wave is the distance from the resting position to a crest or a trough.

Amplitude and Energy of Transverse Waves

You produce a transverse wave on a rope when you move the rope up and down. For a transverse wave, the greatest distance a particle moves from the rest position is to the top of a crest or to the bottom of a trough. This distance is the amplitude of a transverse wave, as shown in **Figure 7**. The energy carried by a transverse wave increases as the amplitude of the wave increases as shown in **Figure 8**.

Amplitude and Energy of Longitudinal Waves

The amplitude of a longitudinal wave depends on the distance between particles in the compressions and rarefactions. When the amplitude of a longitudinal wave increases, the particles in the medium get closer together in the compressions and farther apart in the rarefactions, as shown in **Figure 9** on the next page. In a longitudinal wave, you transfer more energy when you push and pull the end of the spring a greater distance. Just as for transverse waves, the energy carried by a longitudinal wave increases as its amplitude increases.

Reading Check When comparing two longitudinal waves that are traveling through the same medium, how can you tell which has the greater amount of energy?

Figure 8 The wave with the larger amplitude carries more energy and makes the ball bounce higher. ▼

Visual Check What is the source of energy for these waves?

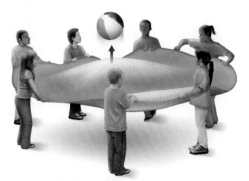
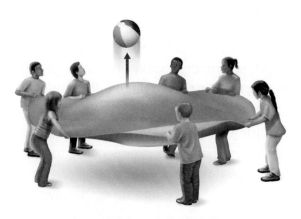

Lower Amplitude Wave
The parachute transfers less energy to the ball.

Higher Amplitude Wave
The parachute transfers more energy to the ball.

Lower-Amplitude Wave

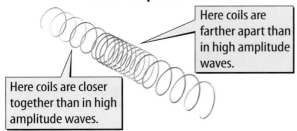

Here coils are farther apart than in high amplitude waves.

Here coils are closer together than in high amplitude waves.

Higher-Amplitude Wave

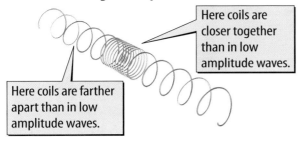

Here coils are closer together than in low amplitude waves.

Here coils are farther apart than in low amplitude waves.

▲ **Figure 9** Amplitude depends on the spacings in the compressions and rarefactions.

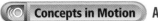 Animation

Wavelength

The **wavelength** *of a wave is the distance from one point on a wave to the same point on the next wave.* The wavelengths of a transverse wave and a longitudinal wave are shown in **Figure 10**. To measure the wavelength of a transverse wave, you can measure the distance from one crest to the next crest or from one trough to the next trough. To measure the wavelength of a longitudinal wave, you can measure the distance from one compression to the next compression or from one rarefaction to the next rarefaction. Wavelength is measured in units of distance, such as meters.

Figure 10 Wavelength is the distance from one point on a wave to the nearest point just like it. ▼

Inquiry MiniLab   20 minutes

How are wavelength and frequency related?

Waves traveling in a material can have different frequencies and wavelengths. Is there a relationship between the wavelength and frequency of a wave?

1. Read and complete a lab safety form.
2. With a partner, stretch a **piece of rope,** approximately 2–3 m long, across a lab table or the floor. Move your hand side to side while your partner holds the other end of the rope in place. Observe the wavelength.
3. Move your hand side to side faster. Observe the wavelength.

Analyze and Conclude

1. **Explain** When was the frequency of the wave higher? Lower?
2. **Key Concept** How are wavelength and frequency related?

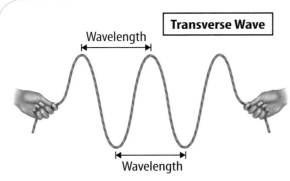

Transverse Wave

Wavelength is the distance from one crest to the next crest or from one trough to the next trough.

Longitudinal Wave

Wavelength is the distance from one compression to the next compression or from one rarefaction to the next rarefaction.

Frequency and Wavelength

- Longer wavelength
- Lower frequency
- One complete wave passes in four seconds.

- Shorter wavelength
- Higher frequency
- Two complete waves pass in four seconds.

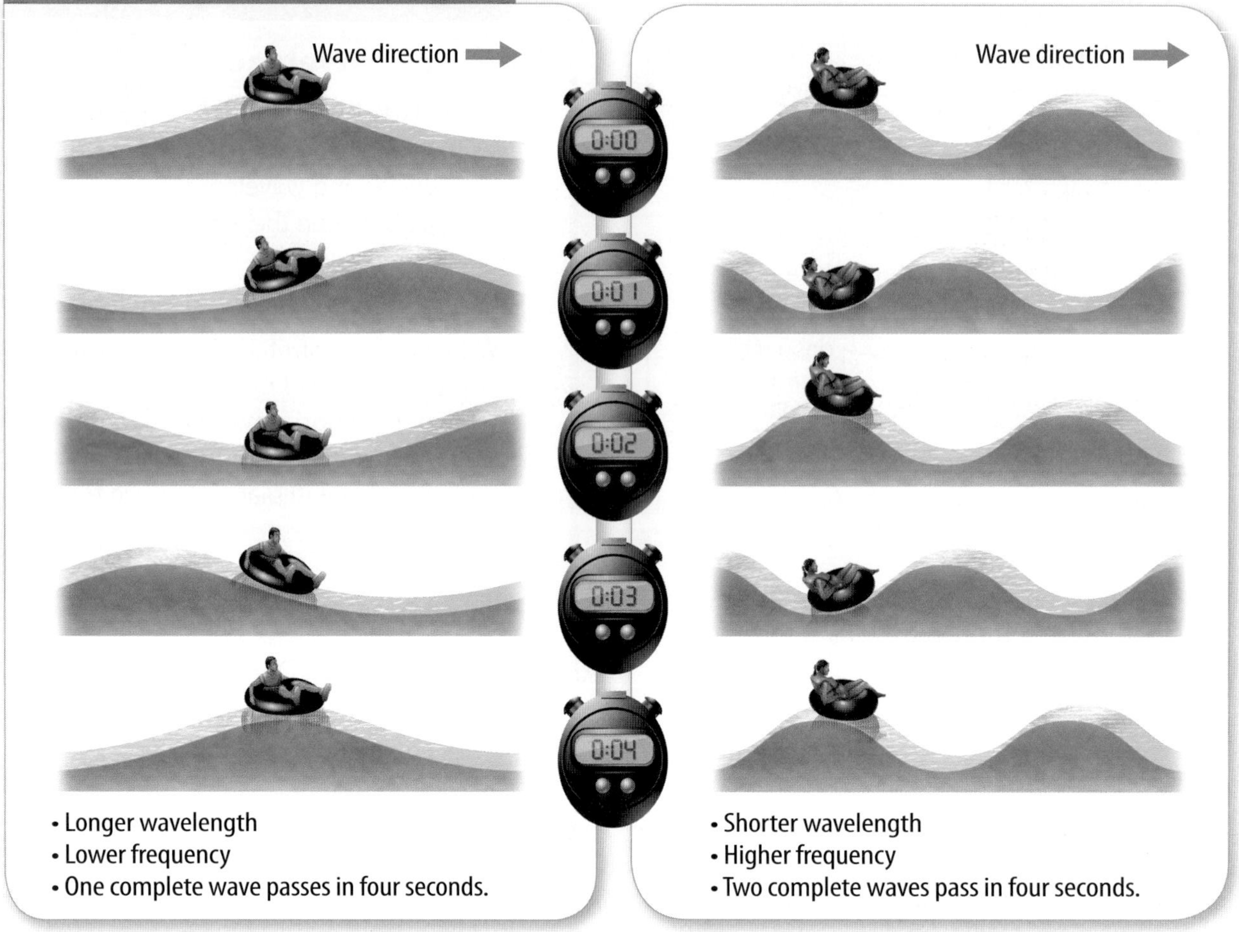

Figure 11 Frequency is the number of wavelengths that pass a point each second. When the frequency increases, the wavelength decreases.

Frequency

Waves have another property called frequency. *The **frequency** of a wave is the number of wavelengths that pass by a point each second.* Frequency is related to how rapidly the object or material producing the wave vibrates. Each vibration of the object produces one wavelength. The frequency of a wave is the same as the number of vibrations the vibrating object makes each second.

Key Concept Check What are three properties of waves?

The Unit for Frequency

The SI unit for frequency is hertz (Hz). A wave with a frequency of 2 Hz means that two wavelengths pass a point each second. The unit Hz is the same unit as 1/s.

Wavelength and Frequency

Figure 11 shows how frequency and wavelength are related. The wavelength of the waves in the left column is longer than that of the wave in the right column. To calculate the frequency of waves, divide the number of wavelengths by the time. For the wave on the left, the frequency is 1 wavelength divided by 4 s, which is 0.25 Hz. The wave on the right has a frequency of 2 wavelengths divided by 4 s, which is 0.5 Hz. The wave on the right has a shorter wavelength and a higher frequency. As the frequency of a wave increases, the wavelength decreases.

Key Concept Check How does the wavelength change if the frequency of a wave decreases? What if the frequency increases?

Wave Speed

Different types of waves travel at different speeds. For example, light waves from a lightning flash travel almost 1 million times faster than the sound waves you hear as thunder.

Wave Speed Through Different Materials

The same type of waves travel at different speeds in different materials. Mechanical waves, such as sound waves, usually travel fastest in solids and slowest in gases, as shown in **Table 2.** Mechanical waves also usually travel faster as the temperature of the medium increases. Unlike mechanical waves, electromagnetic waves move fastest in empty space and slowest in solids.

 Key Concept Check What does wave speed depend on?

Calculating Wave Speed

You can calculate the speed of a wave by multiplying its wavelength and its frequency together, as shown below. The symbol for wavelength is λ, which is the Greek letter *lambda*.

Wave Speed Equation

wave speed (in m/s) = frequency (in Hz) × wavelength (in m)

$$s = f\lambda$$

When you multiply wavelength and frequency, the result has units of m × Hz. This equals m/s—the unit for speed.

Table 2 Speed of Sound Waves in Different Materials

Material	Wave Speed (m/s)
Gases (0°C)	
Oxygen	316
Dry air	331
Liquids (25°C)	
Ethanol	1,207
Water	1,500
Solids	
Ice	3,850
Aluminum	6,420

Math Skills   Use a Simple Equation

Solve for Wave Speed A mosquito beating its wings produces sound waves with a frequency of 700 Hz and a wavelength of 0.5 m. How fast are the sound waves traveling?

1. **This is what you know:**
   - frequency: $f = 700$ Hz
   - wavelength: $\lambda = 0.5$ m

2. **This is what you need to find:**
   - wave speed: s

3. **Use this formula:**
   - $s = f\lambda$

4. **Substitute:**
   - $s = 700$ Hz × 0.5 m = 350 Hz × m

   the values for f and λ into the formula and multiply

5. **Convert units:**
   - (Hz) × (m) = (1/s) × (m) = m/s

Answer: The wave speed is 350 m/s.

Practice

What is the speed of a wave that has a frequency of 8,500 Hz and a wavelength of 1.5 m?

- Math Practice
- Personal Tutor

Lesson 2 Review

Visual Summary

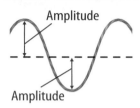

The amplitude of a transverse wave is the maximum distance that the wave moves from its rest position.

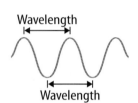

The wavelength of a transverse wave is the distance from one point on a wave to the same point on the next wave, such as from crest to crest or from trough to trough.

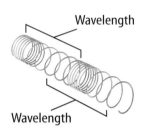

The wavelength of a longitudinal wave is the distance from one point on a wave to the nearest point just like it, such as from compression to compression or from rarefaction to rarefaction.

FOLDABLES

Use your lesson Foldable to review the lesson. Save your Foldable for the project at the end of the chapter.

What do you think NOW?

You first read the statements below at the beginning of the chapter.

3. Waves that carry more energy cause particles in a material to move a greater distance.

4. Sound waves travel fastest in gases, such as those in the air.

Did you change your mind about whether you agree or disagree with the statements? Rewrite any false statements to make them true.

Use Vocabulary

1. For a transverse wave, the _____ depends on the distance from the rest position to a crest or a trough.

2. The unit for the _____ of a wave is the Hz, which means "per second."

Understand Key Concepts

3. **Compare** Which wave would have the greatest wave speed, a wave from a vibrating piano string in an auditorium or a sound wave created by a boat anchor striking an underwater rock? Explain.

4. In which medium would an electromagnetic wave travel the fastest?
   A. air
   B. granite
   C. vacuum
   D. water

Interpret Graphics

5. **Determine** which wave carries the greater amount of energy. Explain.

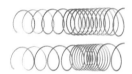

6. **Determine Cause and Effect** Copy and fill in the graphic organizer below.

Critical Thinking

7. **Infer** A loudspeaker produces sound waves that change in wavelength from 1.0 m to 1.5 m. If the wave speed is constant, how did the vibration of the loudspeaker change? Explain.

Math Skills

— Math Practice —

8. **Use a Simple Equation** A water wave has a frequency of 10 Hz and a wavelength of 150 m. What is the wave speed?

Inquiry Skill Practice | Model
25 minutes

How are the properties of waves related?

Materials

meter tapes (2)

masking tape

coiled spring toy

stopwatch or clock with second hand

Safety

All waves have amplitude, wavelength, and frequency. The properties of a wave are related and also determine the amount of energy the wave carries.

Learn It

Scientists create **models** to study many objects and concepts that are difficult to observe directly. You used a coiled spring toy in previous labs to model a longitudinal wave. In this lab, you will use the same toy to model transverse waves.

Try It

1. Read and complete a lab safety form.
2. With a partner, use masking tape to secure the meter tapes to the floor, creating *x*- and *y*-axes.
3. With your partner, stretch the spring toy across the tape representing the *x*-axis. Generate a transverse wave by moving the toy back and forth on the floor. Try to be as steady and even as possible to generate consistent waves.
4. Using the tape on the *y*-axis, measure and record the amplitude of the wave in your Science Journal.
5. Using the tape representing the *x*-axis, measure the wavelength of the wave. Record your measurement.
6. Using a stopwatch, count and record the number of crests or troughs that cross the *y*-axis in 10 seconds.
7. Repeat steps 3–6 for waves with different properties by moving the toy faster and slower.

Apply It

8. Calculate the frequency of each wave generated.
9. Which wave transferred the most energy? Explain.
10. **Key Concept** What happened to the frequency and wavelength of the waves when you moved the spring toy faster and slower? How are frequency and wavelength related?

Lesson 3

Wave Interactions

Reading Guide

Key Concepts
ESSENTIAL QUESTIONS

- How do waves interact with matter?
- What are reflection, refraction, and diffraction?
- What is interference?

Vocabulary
absorption p. 548
transmission p. 548
reflection p. 548
law of reflection p. 549
refraction p. 550
diffraction p. 550
interference p. 551

 Multilingual eGlossary

 Video Science Video

Inquiry Can waves change?

Have you ever watched two waves bump into each other? If so, you might have noticed that the shapes of the waves changed. How do the shapes of waves change when they interact?

Inquiry Launch Lab

20 minutes

What happens in wave collisions?

You might have seen ripples on a water surface spreading out from different points. As the water waves reach each other, they collide. Do waves change after they collide?

1. Read and complete a lab safety form.
2. Stretch a **metal coiled spring toy** about 30–40 cm between you and a partner.
3. Make a wave by grabbing about five coils at one end and then releasing them. Record your observations in your Science Journal.
4. Make waves at both ends of the spring with your partner. Make waves that appear much different from each other so you can distinguish them easily. Then release them at the same time. Observe and record how each wave moves before, during, and after the collision.

Think About This

1. **Describe** how the two waves moved after the coils were released.
2. **Key Concept** How were the two waves affected by their collision?

Interaction of Waves with Matter

Have you seen photos, like the one shown in **Figure 12**, of objects in space taken with the *Hubble Space Telescope?* The *Hubble* orbits Earth collecting light waves before they enter Earth's atmosphere. Photos taken with the *Hubble* are clearer than photos taken with telescopes on Earth's surface. This is because light waves strike the telescope before they interact with matter in Earth's atmosphere.

Waves interact with matter in several ways. Waves can be reflected by matter or they can change direction when they travel from one material to another. In addition, as waves pass through matter, some of the energy they carry can be transferred to matter. For example, the energy from sound waves can be transferred to soft surfaces, such as the padded walls in movie theaters. Waves also interact with each other. When two different waves overlap, a new wave forms. The new wave has different properties from either original wave.

Figure 12 This *Hubble Space Telescope* photo shows a giant cloud of dust and gas called NGC 3603. Many stars are forming in this cloud.

Waves Interact With Matter

The chrome looks shiny because it reflects most of the light waves that strike it.

The glass absorbs only a small amount of the energy carried by light waves. As a result, light waves pass through the glass.

Black paint absorbs almost all the energy carried by light waves.

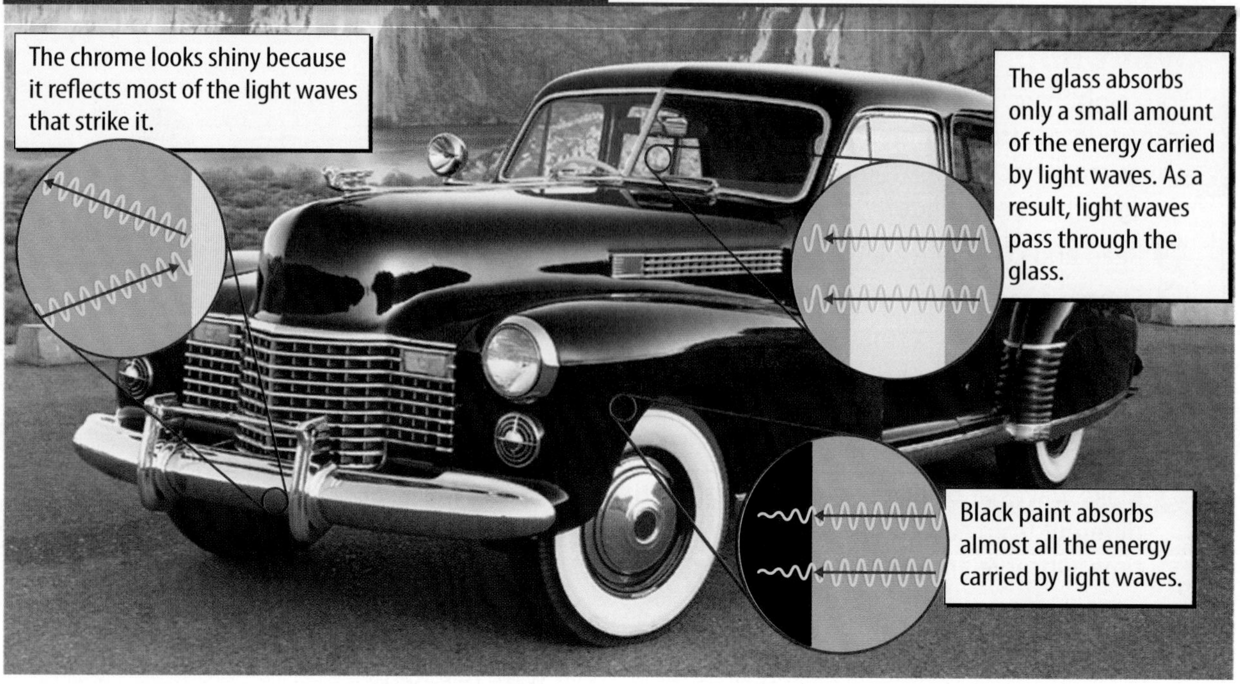

Figure 13 Waves can be absorbed, transmitted, or reflected by matter.

Visual Check Do tires usually absorb, transmit, or reflect light waves? Explain your answer.

Absorption

When you shout, you create sound waves. As these waves travel in air, some of their energy transfers to particles in the air. As a result, the energy the waves carry decreases as they travel through matter. **Absorption** *is the transfer of energy by a wave to the medium through which it travels.* The amount of energy absorbed depends on the type of wave and the material in which it moves.

Reading Check Give one reason why the energy carried by sound waves decreases as those sound waves travel through air.

Absorption also occurs for electromagnetic waves. All materials absorb electromagnetic waves, although some materials absorb more electromagnetic waves than others. Darker materials, such as tinted glass, absorb more visible light waves than lighter materials, such as glass that is not tinted.

Absorption occurs at the surface of the car in **Figure 13**. The car's black paint absorbs much of the energy carried by light waves.

Transmission

Why can you see through a sheet of clear plastic wrap but not through a sheet of black construction paper? When visible light waves reach the paper, almost all their energy is absorbed. As a result, no light waves pass through the paper. However, light waves pass through the plastic wrap because the plastic absorbs only a small amount of the wave's energy. **Transmission** *is the passage of light through an object,* such as the windows in **Figure 13**.

Reflection

When waves reach the surface of materials, they can also be reflected. **Reflection** *is the bouncing of a wave off a surface.* Reflection causes the chrome on the car in **Figure 13** to appear grey instead of black. An object that reflects all visible light appears white, while an object that reflects no visible light appears black.

Key Concept Check What are three ways that waves interact with matter?

All types of waves, including sound waves, light waves, and water waves, can reflect when they hit a surface. Light waves reflect when they reach a mirror. Sound waves reflect when they reach a wall. Reflection causes waves to change direction. When you drop a basketball at an angle, it bounces up at the same angle but in the opposite direction. When waves reflect from a surface, they change direction like a basketball bouncing off a surface.

The Law of Reflection

The direction of a wave that hits a surface and the reflected wave are related. As shown in **Figure 14**, an imaginary line, perpendicular to a surface, is called a **normal**. The angle between the direction of the incoming wave and the normal is the angle of incidence. The angle between the direction of the reflected wave and the normal is the angle of reflection. According to the **law of reflection**, *when a wave is reflected from a surface, the angle of reflection is equal to the angle of incidence.*

 Reading Check What is the law of reflection?

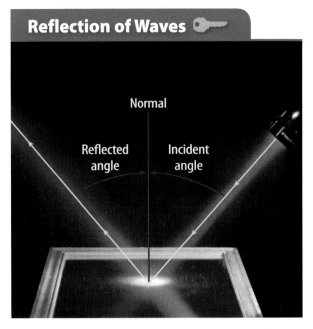

Figure 14 All waves obey the law of reflection. According to the law of reflection, the incident angle equals the reflected angle.

SCIENCE USE V. COMMON USE

normal
Science Use perpendicular to or forming a right angle with a line or plane

Common Use conforming to a standard or common

 **MiniLab** **20 minutes**

How can reflection be used?

Light waves, like all waves, obey the law of reflection. By using mirrors, you can see around corners.

1. Read and complete a lab safety form.
2. Place a **small object** on a table and stand a **book** vertically about 30 cm in front of the object.
3. Position a **mirror** vertically so an observer on the opposite side of the book from the object can see the object. Use **modeling clay** to prop up the mirror.
4. Use **string** to represent the path light waves travel from the observer to the mirror and then to the object. Draw the outlined path in your Science Journal.
5. Repeat steps 3–4 with two mirrors.

Analyze and Conclude

 Key Concept How could three mirrors be used to see the object behind the book?

Refraction of Waves

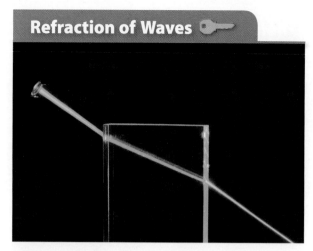

▲ **Figure 15** Refraction occurs when a wave changes speed. The beam of light changes direction because light waves slow down as they move from air into acrylic.

Review Personal Tutor

WORD ORIGIN
refraction
from Latin *refractus*, means "to break up"

Refraction

Sometimes waves change direction even if they are not reflected from a surface. The light beam in **Figure 15** changes direction as it travels from air into acrylic and it changes again when the light beam travels from acrylic into air. When light waves slow down or speed up, they change direction. **Refraction** *is the change in direction of a wave that occurs as the wave changes speed when moving from one medium to another.* The greater the change in speed, the more the wave changes direction.

Diffraction

Waves can also change direction as they travel by objects. Have you ever been walking down a hallway and heard people talking in a room before you got to the open door of the room? You heard some of the sound waves because they changed direction and spread out as they traveled through the doorway.

What is diffraction?

The change in direction of a wave when it travels by the edge of an object or through an opening is called **diffraction.** Examples of diffraction are shown in **Figure 16.** Diffraction causes the water waves to travel around the edges of the object and to spread out after they travel through the opening. More diffraction occurs as the size of the object or opening becomes similar in size to the wavelength of the wave.

Figure 16 Waves diffract as they pass by an object or pass through an opening. ▼

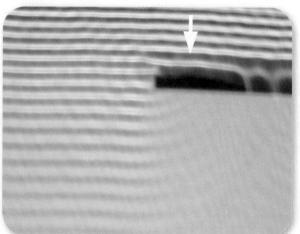
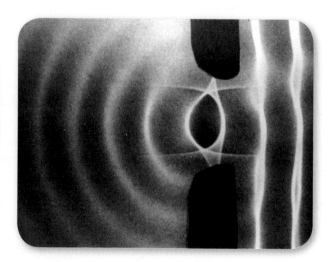

Diffraction of Sound Waves and Light Waves

The wavelengths of sound waves are similar in size to many common objects. Because of this size similarity, you often hear sound from sources that you can't see. For example, the wavelengths of sound waves are roughly the same size as the width of the doorway. Therefore, sound waves spread out as they travel through the doorway. The wavelengths of light waves are more than a million times smaller than the width of a doorway. As a result, light waves do not spread out as they travel through the doorway. Because the wavelengths of light waves are so much smaller than sound waves, you can't see into the room until you reach the doorway. However, you can hear the sounds much sooner.

 Key Concept Check Compare and contrast reflection, refraction, and diffraction.

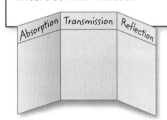

FOLDABLES
Make a tri-fold book and label the columns as shown. Use your book to record information about how waves interact with matter.

Absorption | Transmission | Reflection

Interference

Waves not only interact with matter. They also interact with each other. Suppose you throw two pebbles into a pond. Waves spread out from the impact of each pebble and move toward each other. When the waves meet, they overlap for a while as they travel through each other. **Interference** *occurs when waves that overlap combine, forming a new wave,* as shown in **Figure 17.** However, after the waves travel through each other, they keep moving without having been changed.

Wave Interference

Two waves approach each other from opposite directions.

Wave A → ← Wave B

The waves interfere with each other and form a large amplitude wave.

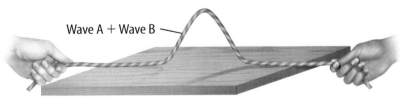

Wave A + Wave B

The waves keep traveling in opposite directions after they move through each other.

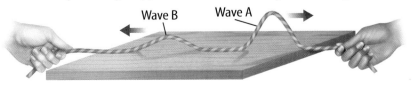

← Wave B   Wave A →

Figure 17 When waves interfere with each other, they create a new wave that has a different amplitude than either original wave.

Visual Check Which wave has the larger amplitude?

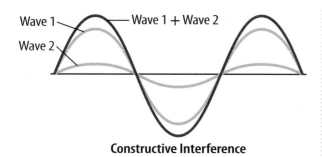

Constructive Interference

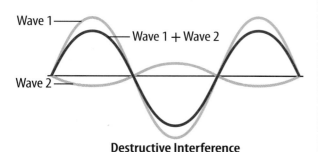

Destructive Interference

▲ **Figure 18** When constructive interference occurs, the new wave has a greater amplitude than either original wave. When destructive interference occurs, the new wave has a smaller amplitude than the sum of the amplitudes of the original waves.

 Animation

ACADEMIC VOCABULARY
constructive
(adjective) pertaining to building or putting parts together to make a whole

Figure 19 A standing wave can occur when two waves with the same wavelength travel in opposite directions and overlap. The wave that forms seems to be standing still. ▼

Constructive and Destructive Interference

As waves travel through each other, sometimes the crests of both waves overlap, as shown in the top image of **Figure 18**. A new wave forms with greater amplitude than either of the original waves. This type of interference is called **constructive** interference. It occurs when crests overlap with crests and troughs overlap with troughs.

Destructive interference occurs when a crest of one wave overlaps the trough of another wave. The new wave that forms has a smaller amplitude than the sum of the amplitudes of the original waves, as shown in the bottom image of **Figure 18**. If the two waves have the same amplitude, they cancel each other when their crests and troughs overlap.

 Key Concept Check Describe two types of wave interference.

Standing Waves

Suppose you shake one end of a rope that has the other end attached to a wall. You create a wave that travels away from you and then reflects off the wall. As the wave you create and the reflected wave interact, interference occurs. For some values of the wavelength, the wave that forms from the combined waves seems to stand still. This wave is called a standing wave. An example is shown in **Figure 19**.

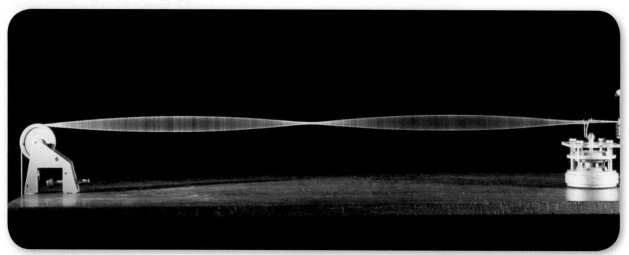

Lesson 3 Review

Assessment — Online Quiz

Visual Summary

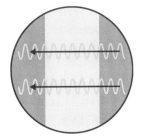

Transmission occurs when waves travel through a material.

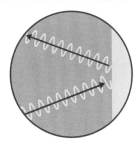

Reflection occurs when waves bounce off the surface of a material.

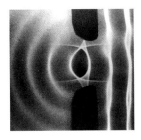

The change in direction of a wave when it travels through an opening is diffraction.

FOLDABLES

Use your lesson Foldable to review the lesson. Save your Foldable for the project at the end of the chapter.

What do you think NOW?

You first read the statements below at the beginning of the chapter.

5. When light waves strike a mirror, they change direction.

6. Light waves travel at the same speed in all materials.

Did you change your mind about whether you agree or disagree with the statements? Rewrite any false statements to make them true.

Use Vocabulary

1 Explain the law of reflection.

2 Distinguish between refraction and diffraction.

Understand Key Concepts

3 Contrast the behavior of a water wave that travels by a stone barrier to a sound wave that travels through a door.

4 Which will NOT occur when a light ray interacts with a smooth pane of glass?
   A. absorption   C. reflection
   B. diffraction   D. transmission

Interpret Graphics

5 Describe what is occurring in the figure below.

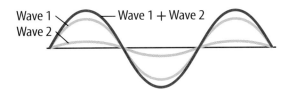

6 Organize Copy and fill in the graphic organizer below. In each oval, list something that can happen to a wave when it interacts with matter.

Critical Thinking

7 Construct Biologists know that chlorophyll, the pigment responsible for photosynthesis in plants, absorbs red light. Design a machine that you could use to test for the presence of chlorophyll.

8 Recommend An architect wants to design a conference room that reduces noise coming from outside the room. Suggest some design features that should be considered in this project.

Inquiry Lab

55 minutes

Materials

meter tapes (2)

masking tape

coiled spring toy

twine, 0.25 m

stopwatch or clock with second hand

Safety

Measuring Wave Speed

When you make a wave on a spring toy, the frequency is how many wavelengths pass a point per second. Wavelength is the distance between one point on the wave and the nearest point just like it. If you can measure the frequency and wavelength of a wave, you can determine the wave speed.

Ask a Question

How can you determine the speed of a wave?

Make Observations

1. Read and complete a lab safety form.
2. Lay the meter tapes on the floor perpendicular to each other to make an x- and y-axis. Fasten them in place with masking tape.
3. Tie a piece of twine around the last coil of the spring toy.
4. With a partner, stretch the spring toy along the x-axis. One person should hold one end at the y-axis. The other person should hold the twine at the end of the outstretched spring.
5. One student creates a transverse wave by moving his or her hand up and down along the y-axis at a constant rate. When the wave is consistent, another student times a 10-second period while the third person counts the number of vibrations in 10 seconds. Record the number of vibrations in your Science Journal in a data table like the one shown below.
6. As the student continues making the wave, another student should estimate the wavelength along the x-axis using the meter tape.
7. Calculate the frequency of the wave. Then calculate the wave speed using the equation, wave speed = frequency × wavelength.
8. Repeat steps 5 through 7 using a different frequency.

Trial	Number of Vibrations in 10 s	Frequency (Hz)	Wavelength (cm)	Wave Speed (cm/s)
1				
2				

Form a Hypothesis

9. Form a hypothesis about the relationship between frequency and wavelength.

Test Your Hypothesis

10. Choose a frequency that you did not use during **Make Observations.** Predict the wavelength for a wave with this frequency.

11. Practice making a wave on the toy spring with your chosen frequency. Repeat steps 4–7 for this wave. Did your prediction of wavelength support your hypothesis? If not, revise your hypothesis and repeat steps 4–7.

Analyze and Conclude

12. **Conclude** How did your prediction of wavelength compare to your measurement?

13. **Think Critically** What measurements were the most difficult to make accurately? Suggest ways to improve on the method.

14. **The Big Idea** How did the wavelength, frequency, and wave speed change for the different waves that you created?

Lab Tips

- ☑ Keep the amplitude constant by moving the same distance on the y-axis in each vibration.
- ☑ Twenty vibrations in 10 s make a wave with a frequency of 2 Hz.

Communicate Your Results

Write a report explaining the steps you took in this lab. Include a table of the measurements you made. Be sure to describe sources of error in your measurements and ways that you might improve the accuracy of your experiment.

Inquiry Extension

Try measuring the wave speed of other waves. Try stretching your spring toy to different lengths or try measuring the wave speed of longitudinal waves. You also might try working with ropes of different thicknesses, different spring toys, or even water in a wave tank.

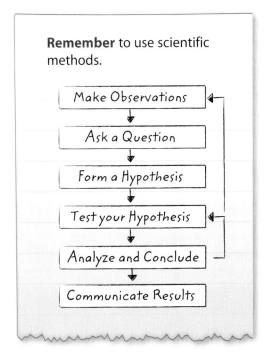

Remember to use scientific methods.

- Make Observations
- Ask a Question
- Form a Hypothesis
- Test your Hypothesis
- Analyze and Conclude
- Communicate Results

Chapter 15 Study Guide

 Waves transfer energy but not matter as they travel.

Key Concepts Summary

Lesson 1: What are waves?
- Vibrations cause **waves**.
- **Transverse waves** make particles in a **medium** move at right angles to the direction that the wave travels. **Longitudinal waves** make particles in a medium move parallel to the direction that the wave travels.
- **Mechanical waves** cannot move through empty space, but **electromagnetic waves** can.

Lesson 2: Wave Properties
- All waves have the properties of **amplitude, wavelength,** and **frequency**.
- Increasing the frequency of a wave decreases the wavelength, and decreasing the frequency increases the wavelength.
- The speed of a wave depends on the type of material in which it is moving and the temperature of the material.

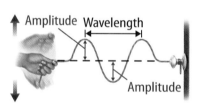

Lesson 3: Wave Interactions
- When waves interact with matter, **absorption** and **transmission** can occur.
- Waves change direction as they interact with matter when **reflection, refraction,** or **diffraction** occurs.
- **Interference** occurs when waves that overlap combine to form a new wave.

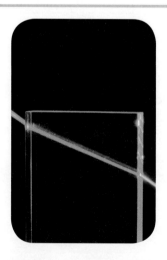

Vocabulary

Lesson 1:
wave p. 529
mechanical wave p. 531
medium p. 531
transverse wave p. 531
crest p. 531
trough p. 531
longitudinal wave p. 532
compression p. 532
rarefaction p. 532
electromagnetic wave p. 535

Lesson 2:
amplitude p. 539
wavelength p. 541
frequency p. 542

Lesson 3:
absorption p. 548
transmission p. 548
reflection p. 548
law of reflection p. 549
refraction p. 550
diffraction p. 550
interference p. 551

Study Guide

- Personal Tutor
- Vocabulary eGames
- Vocabulary eFlashcards

FOLDABLES Chapter Project

Assemble your lesson Foldables as shown to make a Chapter Project. Use the project to review what you have learned in this chapter.

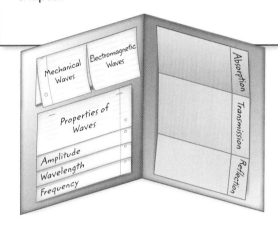

Use Vocabulary

1. A material though which a wave travels is a(n) _____.

2. A(n) _____ is a region where matter is more closely spaced in a longitudinal wave.

3. The Sun gives off energy that travels through space in the form of _____.

4. The product of _____ and wavelength is the speed of the wave.

5. _____ is a property of waves that is measured in hertz.

6. The highest point on a transverse wave is a(n) _____.

7. _____ is when two waves pass through each other and keep going.

Link Vocabulary and Key Concepts

Concepts in Motion — Interactive Concept Map

Copy this concept map, and then use vocabulary terms from the previous page to complete the concept map.

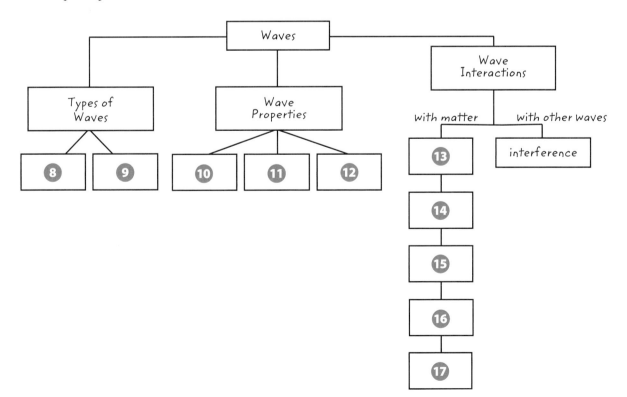

Chapter 15 Study Guide • 557

Chapter 15 Review

Understand Key Concepts

1. What is transferred by a radio wave?
   A. air
   B. energy
   C. matter
   D. space

2. In a longitudinal wave, where are the particles most spread out?
   A. compression
   B. crest
   C. rarefaction
   D. trough

3. Which would produce mechanical waves?
   A. burning a candle
   B. hitting a wall with a hammer
   C. turning on a flashlight
   D. tying a rope to a doorknob

4. Which is an electromagnetic wave?
   A. a flag waving in the wind
   B. a vibrating guitar string
   C. the changes in the air that result from blowing a horn
   D. the waves that heat a cup of water in a microwave oven

5. Identify the crest of the wave in the illustration below.

   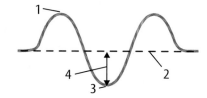

   A. 1
   B. 2
   C. 3
   D. 4

6. If the energy carried by a wave increases, which other wave property also increases?
   A. amplitude
   B. medium
   C. wavelength
   D. wave speed

7. In which medium is the speed of a sound wave the greatest?
   A. air in your classroom
   B. iron railroad track
   C. pool of water
   D. vacuum in space

8. A vibration that produces a wave takes 0.5 seconds to complete. What is the frequency of the wave?
   A. 0.25 Hz
   B. 0.5 Hz
   C. 2 Hz
   D. 4 Hz

9. What does the amount of refraction of a wave depend on?
   A. change in wave speed
   B. location of the normal line
   C. size of the object
   D. size of the opening between objects

10. Two waves travel through each other, and a crest forms with an amplitude smaller than either original wave. What has happened?
    A. constructive interference
    B. destructive interference
    C. reflection
    D. refraction

11. According to the table below, which material is probably a solid?

The Speed of Light in Different Materials	
Material	Speed (km/s)
1	300,000
2	298,600
3	225,000
4	125,000

    A. material 1
    B. material 2
    C. material 3
    D. material 4

558 • Chapter 15 Review

Chapter Review

Assessment — Online Test Practice

Critical Thinking

12 Assess A student sets up a line of dominoes so that each is standing vertically next to another. He then pushes the first one and each falls down in succession. How does this demonstration represent a wave? How is it different?

13 Infer In the figure below, suppose wave 1 and wave 2 have the same amplitude. Describe the wave that forms when destructive interference occurs.

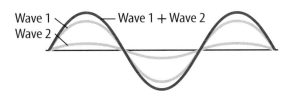

14 Compare A category 5 hurricane has more energy than a category 3 hurricane. Which hurricane will create water waves with greater amplitude? Why?

15 Infer At a baseball game when you are far from the batter, you might see the batter hit the ball before you hear the sound of the bat hitting the ball. Explain why this happens.

16 Evaluate Geologists measure the amplitude of seismic waves using the Richter scale. If an earthquake of 7.3 has a greater amplitude than an earthquake of 4.4, which one carries more energy? Explain your answer.

17 Recommend Some medicines lose their potency when exposed to ultraviolet light. Recommend the type of container in which these medicines should be stored.

18 Explain why the noise level rises in a room full of many talking people.

Writing in Science

19 Write a short essay explaining how an earthquake below the ocean floor can affect the seas near the earthquake area.

REVIEW THE BIG IDEA

20 What are waves and how do they travel? Describe the movement of particles from their resting positions for transverse and longitudinal waves.

21 The photo below shows waves in the ocean. Describe the waves using vocabulary terms from the chapter.

Math Skills

Review — Math Practice

Use Numbers

22 A hummingbird can flap its wings 200 times per second. If the hummingbird produces waves that travel at 340 m/s by flapping its wings, what is the wavelength of these waves?

23 A student did an experiment in which she collected the data shown in the table. What can you conclude about the wave speed and rope diameter in this experiment? What can you conclude about frequency and wavelength?

Wave Speed and Diameter			
Trial	Rope Diameter (cm)	Frequency (Hz)	Wavelength (m)
1	2.0	2.0	8.0
2	2.0	8.0	2.0
3	4.0	2.0	10.0
4	4.0	4.0	5.0

Standardized Test Practice

Record your answers on the answer sheet provided by your teacher or on a sheet of paper.

Multiple Choice

1. Through which medium would sound waves move most slowly?
   A air
   B aluminum
   C glass
   D water

Use the diagram below to answer question 2.

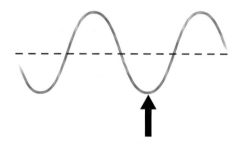

2. The diagram illustrates a mechanical wave. To which does the arrow point?
   A compression
   B crest
   C rarefaction
   D trough

3. Which is an electromagnetic wave?
   A light
   B seismic
   C sound
   D water

4. Which statement about waves is false?
   A Waves transfer matter.
   B Waves can change direction.
   C Waves can interact with each other.
   D Waves can transfer energy to matter.

Use the diagram below to answer question 5.

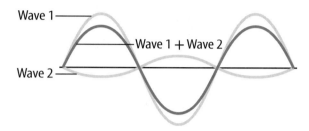

5. Which does the figure illustrate?
   A constructive interference
   B destructive interference
   C diffraction
   D reflection

6. In what region of a longitudinal wave are particles closest together?
   A compression
   B crest
   C rarefaction
   D trough

7. What happens to most of the light waves that strike a transparent pane of glass?
   A absorption
   B diffraction
   C reflection
   D transmission

8. Which wave can travel in both empty space and matter?
   A radio
   B seismic
   C sound
   D water

Standardized Test Practice

Use the figure below to answer question 9.

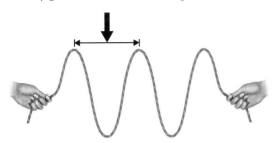

9 To which does the arrow point in the transverse wave diagram?
  A amplitude
  B crest
  C frequency
  D wavelength

10 Which is an example of diffraction?
  A a flashlight beam hitting a mirror
  B a shout crossing a crowded room
  C a sunbeam striking a window
  D a water wave bending around a rock

11 Which property of waves helps explain why a human shout cannot be heard a mile away?
  A absorption
  B diffraction
  C reflection
  D transmission

Constructed Response

Use the table below to answer questions 12 and 13.

	Wave A	Wave B
Number of Wavelengths that pass a point	5	8
Time for wavelengths to pass a point(s)	10	10

12 Which wave has a higher frequency? Why? How does wavelength change as frequency increases?

13 Write and solve an equation to find the speed of wave B if its wavelength is 2 m.

14 Two ocean waves approach a floating beach ball at different times. The second wave has more energy than the first wave. Which wave will have the higher amplitude? Explain your reasoning.

15 Explain why and how the waves from a passing speedboat rock a rowboat. Was the rowboat moved from its original location? Why or why not? Include the definition of a wave in your explanation.

NEED EXTRA HELP?															
If You Missed Question...	1	2	3	4	5	6	7	8	9	10	11	12	13	14	15
Go to Lesson...	2	1	1	1	3	1	3	1	2	3	3	2	2	2	1

Chapter 16

Sound

THE BIG IDEA — How can you produce, describe, and use sound?

Inquiry How does it sound?

Recording a song involves more than just a band playing music. Often, each instrument records alone. Then, a singer records the vocals. Next, an audio engineer blends together the vocals and the sounds from the instruments. Some sounds may need to be adjusted to sound louder, softer, higher, or lower.

- What other properties of sound do you think the soundboard knobs control?
- Why do you think musicians wear headphones?
- How can you produce, describe, and use sound?

Get Ready to Read

What do you think?
Before you read, decide if you agree or disagree with each of these statements. As you read this chapter, see if you change your mind about any of the statements.

1. Sound waves travel fastest through empty space.
2. Hearing loss can be caused by brief, loud sounds, such as a firecracker.
3. Sound waves from different sources can affect one another when they meet.
4. You can tell whether an ambulance is moving toward or away from you by listening to changes in the sound of the siren.
5. Animals use sounds that humans cannot hear to locate objects around them.
6. The method that ships use to locate underwater objects can be used to locate organs or tumors inside the human body.

 Your one-stop online resource

connectED.mcgraw-hill.com

- Video
- WebQuest
- Audio
- Assessment
- Review
- Concepts in Motion
- Inquiry
- Multilingual eGlossary

Lesson 1

Reading Guide

Key Concepts
ESSENTIAL QUESTIONS

- How is sound produced?
- How does sound move from one place to another?
- Why does sound travel at different speeds through various materials?
- What are the functions of the different parts of the human ear?

Vocabulary

sound wave p. 565
longitudinal wave p. 565
vibration p. 565
medium p. 566
compression p. 566
rarefaction p. 566

 Multilingual eGlossary

 Video Science Video

Producing and Detecting Sound

Inquiry What is he listening to?

Think about the different ways people use sounds. Like the boy in the picture, many people use headphones to listen to their favorite songs. Others enjoy attending concerts or playing musical instruments. The sound of a car horn or a person shouting can alert people to danger. How are the different types of sounds produced? How are your ears able to detect the sounds?

Inquiry Launch Lab

10 minutes

What causes sound?

Sound travels through air as vibrations. When those vibrations reach the ear, sound is heard.

1. Read and complete a lab safety form.
2. Stretch **waxed paper** over the top of a **beaker.** Wrap a **rubber band** around the top to hold it tight.
3. Strike the center of the waxed paper gently with the eraser end of a **pencil.** Then strike it harder. How did the sound change? Write your observations in your Science Journal.
4. Sprinkle a few grains of **rice** onto the waxed paper. Strike the paper gently and then harder. Observe how the rice moves each time. Record your observations.

Think About This

1. How was the change in the rice's motion related to the change in sound?
2. **Key Concept** Based on your results, what do you think causes sound?

What is sound?

Everywhere you look, it seems that people have something on their ears! Some are talking on cell phones or listening to music. Some, such as people who work around airplanes, wear ear protection to prevent damage to their hearing. All of these devices have something to do with sound. The sounds you hear are produced by **sound waves**—*longitudinal waves that can only travel through matter.* A **longitudinal wave** *is a wave that makes the particles in the material that carries the wave move back and forth along the direction the wave travels.*

Sources of Sound

Every sound, from the buzzing of a bee to a loud siren, is the result of a vibration. A **vibration** *is a rapid, back-and-forth motion that can occur in solids, liquids, or gases.* The energy carried by a sound wave is caused by vibration. For example, as you pull on a guitar string, you transfer energy to the string. When you let go, the string snaps back and vibrates. As the string vibrates, it collides with nearby air particles. The string transfers energy to these particles. The air particles collide with other air particles and pass on energy, as shown in **Figure 1.**

As the string vibrates, its back-and-forth motion causes a disturbance in the air that carries energy outward from the source of the sound. The disturbance is a sound wave.

Key Concept Check How is sound produced?

Figure 1 As the guitar string vibrates, it transfers energy to nearby air particles.

Visual Check How would the picture of the particles be different if the string vibrated faster?

Figure 2 🗝 Sound waves move away from a source as compressions and rarefactions.

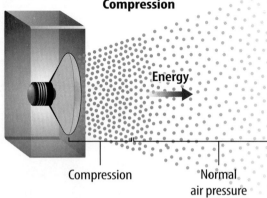
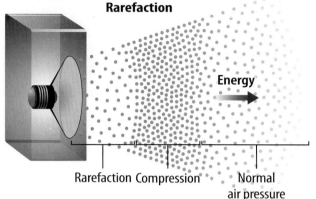

When the speaker cone moves out, it forces particles in the air closer together. This produces a high-pressure area, or compression.

When the speaker cone moves back, it leaves behind an area with fewer particles. This is a low-pressure area called a rarefaction.

SCIENCE USE V. COMMON USE

media
Science Use plural form of *medium*; forms of matter through which sound travels

Common Use a type of mass communication, such as radio

WORD ORIGIN

rarefaction
from Latin *rarefacere*, means "to make rare or less dense"

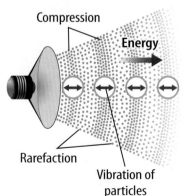

Figure 3 Sound waves carry energy in the direction of the vibrations. ▼

How Sound Waves Travel

Vibrating objects cause sound waves that occur only in matter. For this reason, sound waves must travel through a solid, a liquid, or a gas. *A material in which a wave travels is a* **medium.** You usually hear sound through the medium of air, but sound waves also can travel through other **media,** such as water, wood, and metal. Sound waves cannot travel through empty space because there is no medium to carry the energy.

From a Sound Source to Your Ear If you touch a speaker, like the one in **Figure 2,** you can feel it vibrate as it produces sound waves. Air particles fill a room. Each time the speaker cone moves forward, it pushes air particles ahead of it in the room. This push forces the particles closer together, increasing air pressure in that area. *A region of a longitudinal wave where the particles in the medium are closest together is a* **compression.** With each vibration, the speaker cone moves forward and then back. This motion leaves behind a low-pressure region with fewer air particles, as shown on the right in **Figure 2.** *A* **rarefaction** (rayr uh FAK shun) *is a region of a longitudinal wave where the particles are farthest apart.*

Energy in Sound Waves Suppose you are in the ticket line at the movies. Someone at the back of the line bumps into the next person in line. That person stumbles and bumps the next person before returning to his or her place in line. The energy of the bump continues down the line as each person bumps the next. In the same way, particles of a medium vibrate back and forth as a sound wave carries energy away from a source. This process is shown in **Figure 3.**

✓ **Key Concept Check** How do sound waves travel?

Speed of Sound

You are swimming in a pool when someone taps on the side nearby. Would you hear the sound if your head was under water? Yes, and you would probably hear the sound better than if your head were above water. Sound waves travel faster in water than in air. **Table 1** compares the speed of sound in different media.

Material Two factors that affect the speed of sound waves are the density and the stiffness of the material, or medium. Density is how closely the particles of a medium are packed. Gas particles are far apart and do not collide as often as do the particles of a liquid or a solid. Therefore, sound energy transfers more slowly in a gas than in a liquid or a solid. Notice in **Table 1** that sound travels fastest in solids. In a stiff or rigid solid where particles are packed very close together, the particles collide and transfer energy very quickly.

Sound waves also travel faster in seawater than in freshwater. Seawater contains dissolved salts and has a higher density than freshwater. The fin whale in **Figure 4** emits sounds heard by other whales hundreds of kilometers away.

Table 1 The Speed of Sound	
Material	Speed (m/s)
Air (0°C)	331
Air (20°C)	343
Water (20°C)	1,481
Water (0°C)	1,500
Seawater (25°C)	1,533
Ice (0°C)	3,500
Iron	5,130
Glass	5,640

Table 1 The speed at which sound waves travels through different materials depends on factors such as density and temperature.

◀ **Figure 4** The low-pulse sounds of a fin whale travel more than four times faster through water than they would through air.

Review Personal Tutor

Temperature The temperature of a medium also affects the speed of sound. As the temperature of a gas increases, the particles move faster and collide more often. This increase in collisions transfers more energy in less time. Notice that sound waves travel faster in air at 20°C than in air at 0°C.

In liquids and solids, temperature has the opposite effect. Why do sound waves travel faster in water at 0°C than in water at 20°C? As water cools, the molecules move closer together, so they collide more often. Sound waves travel even faster when the water freezes into ice because it is rigid.

Key Concept Check Why does sound travel at different speeds through various materials?

FOLDABLES
Make a vertical three-tab book and label it as shown. Use it to organize your notes about sound.

Lesson 1
EXPLAIN

▲ Figure 5 The large ears of the fennec fox can detect prey moving underground.

Figure 6 Each part of the ear has an important job that helps you hear. ▼

Visual Check How does the eardrum help you hear?

Detecting Sound

How do you think having large ears helps the fennec fox in **Figure 5**? Sound waves fill the air, and the large outer ear helps funnel sound waves to the inner ear, where sound is detected. Ears also can help you determine the direction a sound comes from. With its large ears, the fennec fox is better able to hear predators approach from greater distances.

The Human Ear

Have you ever cupped your hand around the back of your ear so you could better hear? Why does that work? The human ear has three main parts. The outer ear collects sound waves. By cupping your hand around your ear, you extend the outer ear and therefore collect more waves. The middle ear amplifies sound. The inner ear sends signals about sound to the brain. As shown in **Figure 6**, each part of the ear has a special shape with different parts that help it perform its function.

Key Concept Check What are the functions of the different parts of the human ear?

Functions of Human Ear Parts

Concepts in Motion Animation

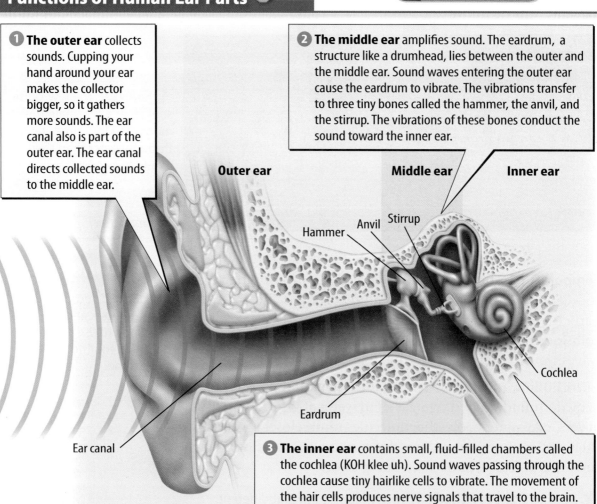

① **The outer ear** collects sounds. Cupping your hand around your ear makes the collector bigger, so it gathers more sounds. The ear canal also is part of the outer ear. The ear canal directs collected sounds to the middle ear.

② **The middle ear** amplifies sound. The eardrum, a structure like a drumhead, lies between the outer and the middle ear. Sound waves entering the outer ear cause the eardrum to vibrate. The vibrations transfer to three tiny bones called the hammer, the anvil, and the stirrup. The vibrations of these bones conduct the sound toward the inner ear.

③ **The inner ear** contains small, fluid-filled chambers called the cochlea (KOH klee uh). Sound waves passing through the cochlea cause tiny hairlike cells to vibrate. The movement of the hair cells produces nerve signals that travel to the brain. The brain interprets these signals as sound.

Inquiry MiniLab

10 minutes

How do you know a sound's direction?

One way an animal can determine the location of a predator is by listening for the sounds the predator makes as it moves. How can ears help determine the direction of sound?

1. Read and complete a lab safety form.

2. Hold the ends of a 1-m piece of **flexible tubing** over your ears. Have your partner use a **pencil** to tap the tubing close to one ear and then the other. Next, have your partner tap at the midpoint of the tubing. Did you notice a change in the sound? Record your observations in your Science Journal.

3. Close your eyes. Have your partner tap at random places along the tubing. Try to guess whether each tap was closest to your left ear, your right ear, or in the middle of the tubing.

4. Switch roles, and tap the tubing as your partner holds the ends to his or her ears.

Analyze and Conclude

1. **Compare** the sounds you heard when your partner tapped the tubing at different locations.

2. **Key Concept** Why is it useful to have two ears instead of one?

Hearing Loss

The harder you pound on a drum, the farther the drumhead travels as it vibrates. What would a drum sound like if the drumhead had a big tear in it? Like a drumhead, the eardrum can be damaged. The eardrum vibrates as pressure changes in the ear. The louder the sound, the farther the eardrum moves in and out as it vibrates.

A sudden loud sound can make the eardrum vibrate so hard that it tears. Damage to the eardrum can cause hearing loss. Also, the tear can allow bacteria into the ear, causing infection. The tear may heal, but thick, uneven scar tissue can make the eardrum less sensitive to sounds.

Listening to loud music over a long period of time also can damage the ears. Look again at **Figure 6** and locate the cochlea. Infection or loud sounds can damage tiny hair cells in the cochlea. Cells that are damaged or die do not grow back, so hearing becomes less sensitive. Many people who work around loud machines, construction, or traffic wear ear protection to prevent damage, as shown in **Figure 7.** Wearing a headset while listening to loud music, however, traps the pressure changes in the ear; this can lead to permanent hearing loss.

Reading Check What are some things that might happen to your ear to cause hearing loss?

Figure 7 This woman is protecting two of her most important senses—sight and hearing.

Lesson 1 Review

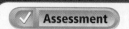 Assessment Online Quiz

Visual Summary

Sound waves carry energy away from a sound source.

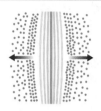
Sound waves move out from a source as a series of compressions and rarefactions.

Ears detect vibrations and interpret them as different sounds.

FOLDABLES

Use your lesson Foldable to review the lesson. Save your Foldable for the project at the end of the chapter.

What do you think NOW?

You first read the statements below at the beginning of the chapter.

1. Sound waves travel fastest through empty space.
2. Hearing loss can be caused by brief, loud sounds, such as a firecracker.

Did you change your mind about whether you agree or disagree with the statements? Rewrite any false statements to make them true.

Use Vocabulary

1. **Define** *longitudinal wave* in your own words.
2. **Identify** the region in a sound wave where the particles are farthest apart.
3. The energy carried by a sound wave comes from a(n) _____ in a medium.

Understand Key Concepts

4. Through which medium would sound travel fastest?
   A. air
   B. iron
   C. cold water
   D. warm water
5. **Summarize** how you produce sound when you tap a pencil against your desk.
6. **Compare and contrast** the functions of the three main parts of the human ear.
7. **Describe** the motion of an air particle as a sound wave passes through it.

Interpret Graphics

8. **Identify** the regions labeled *A* and *B* in the image below. What causes these regions?

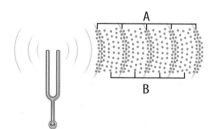

9. **Sequence** Copy and fill in a graphic organizer like the one below that identifies the path of a sound wave as it enters the ear.

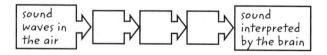

Critical Thinking

10. **Evaluate** A spaceship in a science fiction movie explodes. People in a nearby spaceship hear a loud sound. Is this realistic? Explain.

Cochlear Implants

HOW IT WORKS

Helping Damaged Ears Hear Again

Is there any way to hear again after hair cells in your cochlea are destroyed? Not long ago, the answer was no. But over the last 20 to 30 years, scientists developed a way to bypass the damaged cells. It is called a cochlear implant—a device that uses electrical signals to stimulate the nerves that go from the ear to the brain. How does it work?

First a surgeon implants the interior part of the device under the scalp and into the inner ear. Then the exterior part of the device is put to work. With hearing restored, you are once again connected to the sounds in the world around you!

1. A microphone receives sound waves from the environment. A speech processor changes sounds into electrical signals.

2. The electrical signals are then sent across the scalp from the transmitter to a receiver.

3. The receiver sends the signals through a wire to electrodes implanted in the cochlea.

4. Nerves in the inner ear pick up the signals and send them to the brain. The brain interprets the signals as sound.

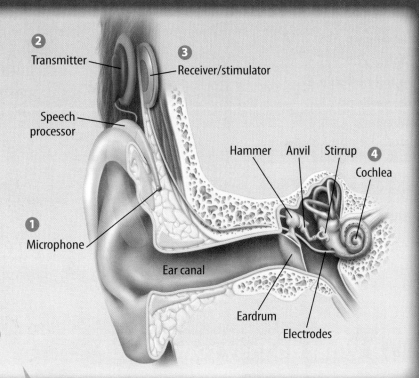

It's Your Turn

MAKE A POSTER How does a hearing aid work? How is it different from a cochlear implant? Research these questions and create a poster from what you discover.

Lesson 1 EXTEND

Lesson 2

Reading Guide

Key Concepts
ESSENTIAL QUESTIONS

- How are amplitude and intensity related to energy?
- What is the relationship among frequency, pitch, and wavelength?
- How can you recognize sounds from different sources?
- In what ways are musical sounds produced?

Vocabulary

amplitude p. 573
intensity p. 574
wavelength p. 575
frequency p. 575
pitch p. 575
Doppler effect p. 576
interference p. 576
resonance p. 578

Multilingual eGlossary

Video Science Video

Properties of Sound Waves

Inquiry What do they sound like?

High in the mountains of Switzerland, you might hear the clear, mellow tones of an alphorn. At one time, herders used the horns to call or soothe cows. How do you think the shape of the horn affects the sound? Why do you think the sound is able to travel so far?

Inquiry Launch Lab

15 minutes

How can sound blow out a candle?

Sound waves carry energy. Can the energy in a sound wave affect a nearby candle flame?

1. Read and complete a lab safety form.
2. Cut off the neck of a **balloon** with **scissors.** Stretch the remaining part of the balloon over the wide end of a **small funnel.**
3. Set a ball of **modeling clay** on a table. Insert a small **candle** in the clay, and use a **safety match** to light it.
4. Hold the funnel with the narrow end pointing toward the lit candle, about 2 cm away.
   ⚠ *Do not let the funnel touch the flame.*
5. Sharply strike the rim of the funnel several times with a **ruler** so that it makes a loud sound. What happens to the candle flame? Record your observations in your Science Journal.
6. Strike the funnel several more times. Each time, vary the amount of energy you use. Record your observations.

Think About This

1. What was different about the flame when you made a soft sound compared to when you made a loud sound? Why do you think this happened?
2. 🔑 **Key Concept** How did the amount of energy you used to strike the funnel affect the sound? How did it affect the flame?

Energy of Sound Waves

Shhhhh! How do you change your voice from a yell to a whisper? The energy a sound wave carries depends on the amount of energy that caused the original vibration. To speak softly, just use less energy!

Amplitude

How do the sound waves produced when you yell and when you whisper differ? The more energy you put into your voice, the farther the air particles move as they vibrate back and forth. *For a longitudinal wave,* **amplitude** *is the maximum distance the particles in a medium move from their rest positions as the wave passes through the medium.* As the energy in a sound wave increases, its amplitude increases. Sound waves with small and large amplitudes are shown in **Figure 8.**

✅ **Key Concept Check** How is the amplitude of sound related to energy?

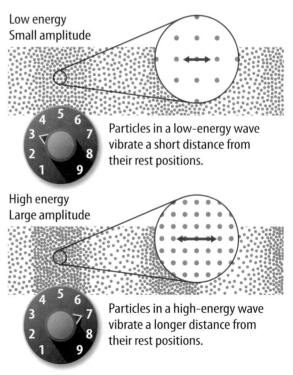

Particles in a low-energy wave vibrate a short distance from their rest positions.

Particles in a high-energy wave vibrate a longer distance from their rest positions.

Figure 8 🔑 Particle spacing differs in high-energy and low-energy waves.

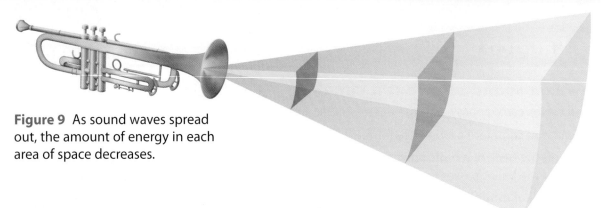

Figure 9 As sound waves spread out, the amount of energy in each area of space decreases.

Amplitude, Intensity and Loudness

Imagine blowing into a trumpet, like the one in **Figure 9**. Sound waves leave the horn with a certain amount of energy and, therefore, a certain amplitude. Recall that amplitude is the distance the particles of air vibrate back and forth. Loudness is how you perceive the energy of a sound wave. So, a wave with a greater amplitude will produce a louder sound.

Why do sounds get quieter as you get farther from the source? Think of what happens as a sound wave travels away from the horn. As the particles of air in front of the horn vibrate back and forth, they collide with, and transfer energy to, surrounding particles of air. As the energy spreads out among more and more particles, the intensity of the wave decreases. **Intensity** *is the amount of sound energy that passes through a square meter of space in one second.*

 Key Concept Check How is the intensity of sound related to energy?

As a sound wave travels farther from the horn, there is a larger area of particles sharing the same amount of energy that left the horn. Therefore, the farther you are from the horn, the less energy passing trough one square meter. This results in less intensity of the wave. As intensity decreases, amplitude decreases. Therefore, loudness decreases.

The Decibel Scale

The unit decibel (dB) describes the intensity and, in turn, the loudness of sound. Decibel levels of common sounds are shown in **Figure 10**. Each increase of 10 dB indicates the sound is about twice as loud and has about 10 times the energy. For example, the decibel level of city traffic is about 85 dB, and the level of a rock concert is about 105 dB. This means a concert, which is 20 dB higher, has about 10 × 10, or 100 times more energy than traffic. Recall that a loud sound can make the eardrum vibrate so hard that it tears. As sounds get louder, the amount of time you can listen without hearing loss gets shorter.

Decibel Scale

Figure 10 The decibel scale rates the loudness of some common sounds.

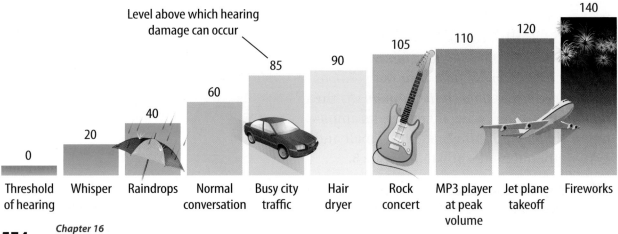

Wavelength and Frequency

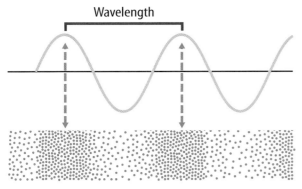
Low frequency; long wavelength

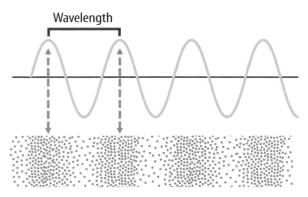
High frequency; short wavelength

Describing Sound Waves

Sounds depend on many properties of the sound waves that enter your ear. Loudness or softness depends on the amplitude of the wave. You also might describe sound according to how frequently the waves occur or how long the waves are.

Wavelength

One property of a sound wave is its wavelength. *The distance between a point on one wave and the nearest point just like it is called* **wavelength**. For example, you could measure a wavelength as the distance between the midpoint of one compression, or rarefaction and the midpoint of the next compression, or rarefaction, as shown in **Figure 11**.

Frequency and Pitch

Suppose you could count sound waves produced by playing middle C on a piano. You would find that 262 wavelengths pass you each second. *The* **frequency** *of sound is the number of wavelengths that pass by a point each second.* Notice in **Figure 11**, that as the wavelength of a sound wave decreases, its frequency increases. The frequency of one vibration, or wavelength, per second is called a hertz (Hz). The frequency of middle C on a piano is 262 Hz.

The perception of how high or low a sound seems is **pitch**. A higher frequency produces a higher pitch. For example, an adult male voice might range from 85 Hz to 155 Hz. An adult female voice might range from about 165 Hz to 255 Hz.

The human ear can detect sounds with frequencies between about 20 Hz and 20,000 Hz. Frequencies above this range are called ultrasound. The range of sounds heard by various animals is shown in **Figure 12**.

Key Concept Check What is the relationship among frequency, pitch, and wavelength?

Figure 11 If you compare two sound waves, the wave with the longer wavelength has a lower frequency. ▼

Visual Check How can you tell that the sounds have the same intensity?

▲ **Figure 12** Many animals can hear sounds outside the range of human hearing.

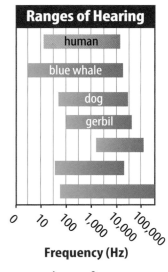

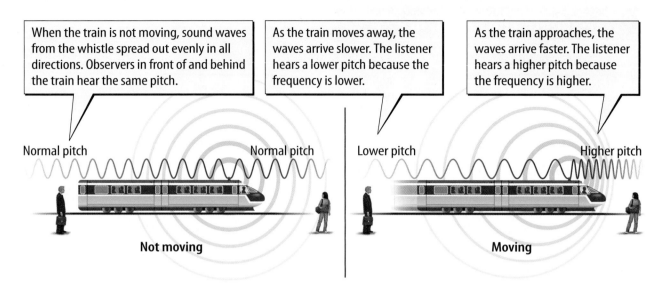

▲ Figure 13 Sound waves bunch together ahead of a moving source.

The Doppler Effect

You might have heard the high pitch of a train whistle as the train approaches. As the train passes, the pitch drops. Sound frequency depends on the motions of the source and the listener. Compare the wave frequencies in front of and behind the moving train in **Figure 13.** The frequency increases if the distance between the listener and the source is decreasing. The frequency decreases if the distance between the listener and the source increases. *The change of pitch when a sound source is moving in relation to an observer is the* **Doppler effect.**

Sound Interference

If you walk through a room with stereo speakers at each end, the sound might seem louder in some places and softer in others. The waves from each speaker interact with one another. **Interference** *occurs when waves that overlap combine, forming a new wave.*

Figure 14 shows why this happens. When compressions meet, they join to form a wave with higher intensity and greater amplitude. This is called constructive interference. However, when a compression meets a rarefaction, the intensity and amplitude decrease. This is destructive interference.

▼ Figure 14 Waves' interference can result in an increase or a decrease in amplitude.

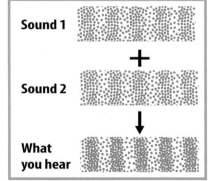

Constructive Interference
When the compressions and rarefactions of waves overlap, the combined compressions have greater intensity.

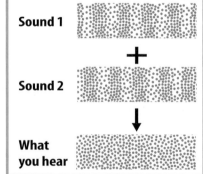

Destructive Interference
When the compressions of one wave overlap the rarefactions of another wave, the waves cancel and the result is no sound.

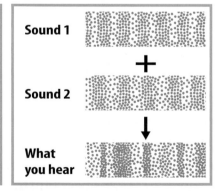

Beats
When the compressions of two waves are slightly offset, a pattern of increasing and decreasing compressions, called beats, occurs.

Inquiry MiniLab

10 minutes

How can you hear beats?

When musicians play tones with slightly different pitches, the combined sound creates a beat.

1. Read and complete a lab safety form.
2. Wrap two **rubber bands** around a **box** so that they stretch across the open side. Listen as you pluck the rubber bands.
3. Loosen or tighten the rubber bands until their pitches are low and very close. Put your ear near the box, and pluck both rubber bands at the same time. Listen carefully for slight variations in the loudness of the combined sound.
4. Repeat several times. Adjust the rubber bands until you hear beats. Record your observations in your Science Journal.

Analyze and Conclude

1. Describe how you produced the beats in the sound.
2. **Key Concept** How do you think musicians playing together can avoid beats?

Beats

Have you ever been to a concert where the musicians start by all playing the same note? Why do they do that? If the pitches of the notes are slightly different, the sounds will interfere. The audience might hear the notes get louder and softer several times a second. The repeating increases and decreases in amplitude are beats. Look back at **Figure 14.** The difference in frequencies determines how often beats occur. If one musician plays a note with a pitch of 392 Hz and another plays a note with a pitch of 395 Hz, the difference is 3 Hz. Beats will occur 3 times each second. Musicians can avoid beats by playing notes at the correct pitch on their instruments.

Fundamental and Overtones

When a musician plucks a guitar string, the string vibrates with a certain sound. If the musician plucks it again in the exact same way, the sound will be the same. All objects tend to vibrate with a certain frequency that depends on the object's properties.

The lowest frequency at which a material naturally vibrates is called its fundamental. Higher frequencies at which the material vibrates are called overtones. Objects vibrate with both a fundamental and overtones, as shown in **Figure 15.** The interference of these waves produces the sound you hear.

Reading Check What is the difference between a fundamental and overtones?

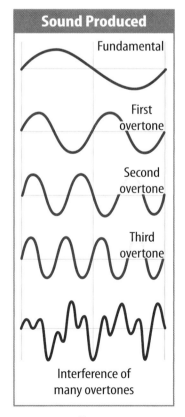

Figure 15 A fundamental and overtones combine to produce an object's sound.

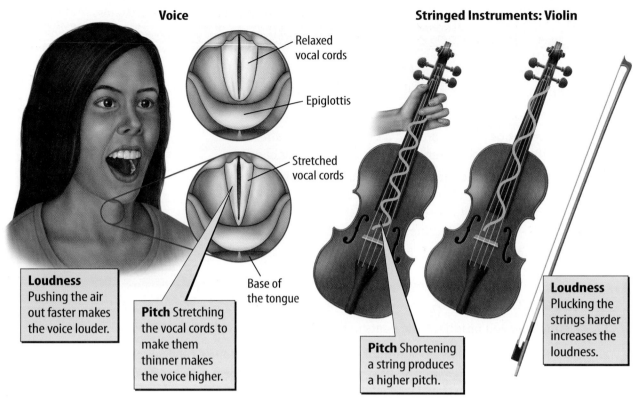

Figure 16 Musicians change the pitch and loudness of sound in different ways.

FOLDABLES

Make a vertical two-tab book and label it as shown. Use it to explain what you have learned about the relationships between sound properties and energy.

WORD ORIGIN

resonance
from Latin *resonare*, means "sound again"

Music

Have you ever been listening to your favorite music and someone tells you "Turn off that noise!" How are music and noise different? Unlike noise, music is sound with a pleasing pattern.

Sound Quality

The unique sound of a musical instrument is a mix of its fundamental and overtones. These waves interact to form a distinct sound, or timbre. Suppose, for example, a clarinet player and a piano player both play the same note. The fundamental of both instruments is the same. The number and intensity of the overtones, however, differs. Overtones produce the complex sound waves that let you distinguish the unique sound quality of each instrument.

 Key Concept Check How can you recognize sounds from different sources?

Resonance

If you hold a guitar string by its ends and have a friend pluck the string, the sound is almost too low to hear. Instruments use resonance to amplify sound. **Resonance** *is an increase in amplitude that occurs when an object vibrating at its natural frequency absorbs energy from a nearby object vibrating at the same frequency.* The vibrating string causes the back of the guitar and the air inside to vibrate by resonance. The sound is then much louder.

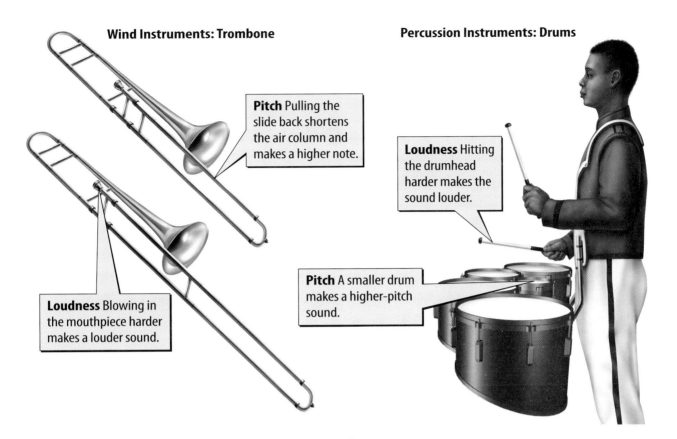

Types of Musical Instruments

As shown in **Figure 16,** different types of instruments control the pitch and the loudness of sound in different ways.

String Instruments Musical instruments that have strings, such as a guitar, a violin, a harp, and a piano, produce sound when the string vibrates. A player plucks the strings of a guitar or a harp. The motion of a bow vibrates the strings of a violin. Pressing a piano key causes a felt-covered hammer to strike a particular string inside the piano. When the string inside the piano vibrates, you hear a tone. The pitch a string makes depends on its length and its thickness.

Pitch also depends on how tightly the string is stretched and the material it is made from. You can hear sounds a stringed instrument makes because of resonance between the string and the instrument's hollow body. Plucking or pressing the strings harder increases the volume.

Wind Instruments The vibrating medium in wind instruments, such as a saxophone or a trumpet, is air. Either your lips or a thin piece of wood, called a reed, vibrates. An air column then vibrates by resonance. The length of the air column determines pitch. Blowing harder increases the volume.

Percussion Instruments You make sound with a percussion instrument by striking it. Examples include a drum, cymbals, and a bell. The pitch depends on the instrument's size, its thickness, and the material from which it is made. Resonance can make the sound of a percussion instrument louder.

Voice The source of sound in your voice is vocal cords. Muscles in your throat allow you to increase the pitch of your voice by pulling the cords tighter. Other parts of your mouth and throat also affect the sounds your voice makes. You can make your voice louder by pushing out the air with a greater force.

Key Concept Check In what ways are musical sounds produced?

Lesson 2 Review

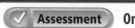

Visual Summary

The intensity of sound decreases farther away from a source.

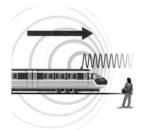

The pitch of sound increases if the distance between a sound source and a listener decreases.

An instrument's sound is a mix of its fundamental and overtones.

FOLDABLES

Use your lesson Foldable to review the lesson. Save your Foldable for the project at the end of the chapter.

What do you think NOW?

You first read the statements below at the beginning of the chapter.

3. Sound waves from different sources can affect one another when they meet.

4. You can tell whether an ambulance is moving toward or away from you by listening to changes in the sound of the siren.

Did you change your mind about whether you agree or disagree with the statements? Rewrite any false statements to make them true.

Use Vocabulary

1 As the _____ of a sound wave increases, the pitch of the sound gets lower.

2 Describe the Doppler effect in your own words.

3 Use the term *resonance* in a complete sentence.

Understand Key Concepts

4 Which increases if the amplitude of a sound wave increases?
   A. intensity   C. pitch
   B. interference   D. sound quality

5 Identify the vibrating media in three different types of musical instruments.

6 Explain how you can recognize the sound of a flute.

7 Analyze If the frequency of a sound wave increases, what happens to the wavelength?

Interpret Graphics

8 Identify Copy and fill in a graphic organizer like the one below. Use it to identify and briefly explain four properties that distinguish sounds.

Critical Thinking

9 Decide If you double the amplitude of a string vibrating at 10 Hz, will it sound louder? Why or why not?

10 Construct a drawing of two waves that represent a violin and a tuba each playing a note with the same frequency and loudness.

Inquiry Skill Practice: Manipulate Variables

40 minutes

How can you use a wind instrument to play music?

Wind instruments use a vibrating column of air to create a sound. By changing the length of the column of air vibrating inside the instrument, you change the frequency of the vibrating air.

Materials

6 test tubes

test-tube rack

beaker

ruler

straw

Safety

Learn It

When you **manipulate variables**, you change only one factor, called the independent variable. A factor that changes as a result of the independent variable is a dependent variable. By changing only one variable, you know what causes any effects you observe.

Try It

1. Read and complete a lab safety form.

2. Place 6 test tubes in a rack. Add just enough water to cover the bottom of the first tube. Then, measure the length of the air column inside the test tube. Record your measurement in your Science Journal.

3. Blow across the top of the tube with a straw. Record your observations about the sound.

4. Add a greater amount of water to another test tube. Observe the difference in the sound as you blow across the tube. Measure the length of the air column, and again record your observations.

5. Think about how the pitch of the sound relates to the length of the air column. Choose a simple song to play by blowing on the test tubes.

6. Add different amounts of water to the test tubes to make different pitches. The pitches should correspond to the notes of your song.

7. Measure and record the air column length for each pitch. Write the pattern of lengths needed for your song.

Apply It

8. Identify the independent and the dependent variables.

9. Describe the relationship between pitch and the length of the air column.

10. **Key Concept** Based on what you have learned from this lab, explain how musicians use wind instruments to play music.

Lesson 2
EXTEND
581

Lesson 3

Reading Guide

Key Concepts
ESSENTIAL QUESTIONS

- In what ways does sound interact with matter?
- How can people control sound?
- What are some ways to use ultrasound?

Vocabulary
absorption p. 584
reflection p. 584
echo p. 584
reverberation p. 585
acoustics p. 585
echolocation p. 587
sonar p. 587

g Multilingual eGlossary

Using Sound Waves

Inquiry Where is it?

These workers are preparing to lower a special device into the water that uses sound to search for underwater objects. Sometimes they locate sunken ships, such as the oil tanker in the large image. What properties of sound waves are useful for finding things under water?

Inquiry Launch Lab

15 minutes

Why didn't you hear the phone ringing?

A cell phone ringing on a countertop can be heard from far away, but the same phone in your jacket pocket is not so easy to hear. What explains this difference?

1. Read and complete a lab safety form.
2. Set a **kitchen timer** for 1 s. Hold it about 15 cm from your ear, and listen to the sound.
3. Locate the exit hole for sound waves on the back of the timer. Set the timer to a 5-s delay. Hold a **foil pie pan** flat against the exit hole. Move your ear about 15 cm from the timer and pie pan. Observe the difference in the sound. Record your observations in your Science Journal.
4. Set the timer with a 5-s delay. Place the timer in a **shoe box,** and cover it with several crumpled **paper towels.** Listen with your ear 15 cm from the box. Record your observations.

Think About This

1. Compare the movement of sound through the foil pan and through air.
2. What do you think changed the sound when you covered the timer with paper towels?
3. **Key Concept** Based on your results, why do you think a cell phone is harder to hear in a jacket pocket than on a countertop?

Sound Waves and Matter

Why do the cheers of a crowd in an indoor gymnasium sound so different from a crowd yelling at an outdoor football game? Sound waves at the football game spread out with few barriers. What happens to sound waves when they strike a different medium, such as the walls of a building?

Transmission

Have you ever heard someone talking in the next room? This is possible because of transmission—the movement of sound waves through a medium. When sound waves move from air into a wall, the vibrations of air particles cause particles in the solid wall to vibrate. Even though solids transmit sound waves better than gases, most sound waves do not move easily from gases into a solid. However, loud sounds, which have a lot of energy, will move into and through the solid wall. Waves of quieter sounds, with less energy, may be partially or completely blocked. These waves don't carry enough energy to cause much vibration in the wall. As the vibrations reach the next room, they transfer the remaining energy to the air particles, and you hear the sound. The amplitude of the sound is lower because the wall could not transmit all of the energy.

FOLDABLES

Make a vertical five-tab book. Label it as shown. Use it to organize your notes about the different ways that sound interacts with matter.

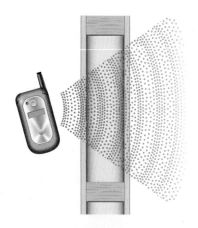

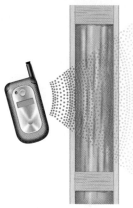

▲ Figure 17 Insulation absorbs much of the energy in a sound wave and converts it to a small amount of heat.

WORD ORIGIN

echo
from Greek *ekhe*, means "sound"

Figure 18 Sound waves reflect from a surface at the same angle at which they strike the surface.

Absorption

If you throw a tennis ball at a pillow, it will not bounce back. Most of the ball's energy goes into the pillow. Some materials, such as the wall insulation in **Figure 17**, act like the pillow when sound waves strike them. *The transfer of energy by a wave to the medium through which it travels is called* **absorption.** How well a material absorbs the energy of a sound wave depends on various factors, such as its inner structure and the amount of air in it. Rather than passing from one particle to another, some of the sound energy changes to heat due to friction.

Reflection

What happens if you throw a tennis ball at a hard surface? The ball probably bounces back at you. Similarly, a sound wave might bounce back when it strikes a different medium. *The bouncing of a wave off a surface is called* **reflection.** The way in which sound waves reflect is shown in **Figure 18**. The angle at which a sound wave strikes a surface is always equal to the angle at which the sound wave is reflected off of the surface.

Key Concept Check What are some ways in which sound waves interact with matter?

Echoes

Have you ever yelled a name in a gym and heard the same voice yell back? That was you, of course! As you yelled, you heard the original sound of your own voice. Then you heard the sound again after the sound waves reflected off the walls of the gymnasium and traveled back to your ears. *A reflected sound wave is an* **echo.**

Sound waves travel at about 343 m/s in air, or 34.3 m in 0.1 s And, the brain holds onto a sound for about 0.1 s. When the reflecting surface is far enough away that the sound waves take more than 0.1 s to return, the listener hears the original sound followed by the reflected sound. So, if you clap your hands at one end of a long room, you hear an echo only if the sound wave returns more than 0.1 s later.

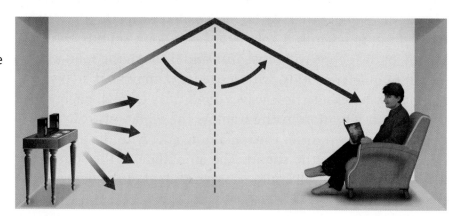

Reverberation

In many closed spaces, sound waves reflect from surfaces that are different distances from the listener. Because some waves travel farther than other waves, reflected waves reach the listener at different times. *The collection of reflected sounds from the surfaces in a closed space is called* **reverberation**.

Sound waves that reach the listener directly are heard sooner than reflected sound waves. If each reflected wave reaches the ear before the previous sound fades, the original sound seems to last longer. However, too much reverberation can make words hard to understand because the echoes of old sounds can interfere with new ones.

Acoustics

In a room with no furniture, rugs, or drapes, the sound waves of footsteps and speech bounce around the room and sound loud. Soft or fuzzy materials, such as the rug, curtains, and padded furniture on the right of **Figure 19**, absorb much of the energy of sound waves. Footsteps are almost silent. Voices are softer. *The study of how sound interacts with structures is called* **acoustics**. Acoustical engineers use their knowledge of sound transmission, absorption, and reflection to control sounds.

Inquiry MiniLab 25 minutes

How fast is sound?

The speed of sound in air is about 343 m/s. You can use echoes to measure this speed.

1. Read and complete a lab safety form.
2. Use a **meterstick** to measure a spot 30 m from a wall. Standing at this spot, clap your hands once and listen for the echo.
3. Clap in time with the echo. If you hear the echo after each clap, clap slightly faster.
4. When your clapping matches the echo speed, have your partner use a **stopwatch** to measure the time as you clap 25 times. Record the time in your Science Journal.
5. Repeat steps 2–4 at distances of 40 m and 50 m from the wall.

Analyze and Conclude

1. **Calculate** the speed of sound for each measurement: *speed = distance ÷ time*.
2. **Key Concept** Describe how echoes helped you measure the speed of sound.

Figure 19 Soft materials reduce the reverberation in a room.

Visual Check Why do you hear reverberation only in a closed space?

Noise Pollution and Control

One way acoustical engineers control sound is by developing methods to protect people from noise pollution. Think about the noises people might hear during a typical day. Trucks rumble by, cars honk their horns, and claps of thunder boom during a storm. At home, you might use noisy appliances such as a hair dryer, a vacuum cleaner, or a dishwasher. Severe noise pollution can result in hearing loss, stress, and other types of health problems. Laws limit the noise that might be produced by machinery or landing aircraft. The government requires ear protection for workers in many jobs.

Noise-canceling earphones, such as those worn by the person in **Figure 20,** work in several ways. They cover the ears and can block incoming sound waves. Other types of earphones reduce the sound by analyzing incoming sound waves and then producing waves that create destructive interference.

Designing Spaces

Acoustical engineers develop ways to control sound in buildings. When you enter a concert hall such as the one in **Figure 21,** you might think that the unusual appearance is just for looks. However, engineers have carefully chosen shapes and materials to control sound waves. The stage has a wooden floor to improve vibrations. The curved panels on the ceiling reflect sound waves in different directions to fill the space. However, the recording studio has foam panels on the walls. These soft materials absorb sound waves to **prevent** reverberation as the musician performs.

Key Concept Check What are some ways in which people can control sound?

▲ **Figure 20** Noise-canceling earphones protect the worker from noises that otherwise would damage his hearing.

ACADEMIC VOCABULARY

prevent
(verb) to keep from happening

Figure 21 Engineers design concert halls and recording studios to control sounds. ▼

Visual Check What methods have designers used to control sound in each of these spaces?

① A dolphin sends out a series of high-pitched clicks.

② When a sound wave strikes an object or another dolphin, some of the sound reflects back.

Figure 22 Organs in the dolphin's head send out ultrasonic waves and analyze the echoes.

Ultrasound

You have read that the waves of ultrasound have a higher frequency than humans can hear. Both animals and humans use these high-frequency sound waves.

Echolocation

Recall that many animals, such as bats and whales, can hear sounds above the range of human hearing. Animals use some of these ultrasound frequencies to communicate. They use other frequencies to locate and identify objects. *The process an animal uses to locate an object by means of reflected sounds is* **echolocation** (e koh loh KAY shun). The dolphins in **Figure 22** might use an echo to determine an object's distance, shape, and speed or to find other dolphins.

Sonar

If you go fishing often, you probably wish you could easily discover exactly where the fish are located. A device called a fish-finder does just that by using a technology similar to echolocation. A fish-finder is an example of **sonar**, *a system that uses the reflection of sound waves to find underwater objects*. The word *sonar* is an acronym for Sound Navigation and Ranging. Ships use sonar to send a high frequency sound wave into the water. As the sound wave moves deeper, it spreads out, forming a cone, or beam. When the sound wave strikes something within this beam, it bounces back to the ship. Sonar contrasts signals that strike the ocean floor with signals from other objects. It measures the amount of time between when the sound wave leaves and when it bounces back. The sonar system then calculates the distance and draws an image on a screen. The picture of the sunken oil tanker on the first page of this lesson shows what a sonar image might look like.

Reading Check How can people "see" under water using sonar?

Math Skills

Use a Formula

You can use a formula to calculate sonar distances. For example, sound travels at **1,530 m/s** in seawater. A ship sends out a signal to find the water's depth. The signal returns in **5.0 s**. How deep is the water?

① Use the speed = distance/time formula to calculate distance.

distance = speed × time

② The signal traveled to the ocean's bottom and back to the ship. Divide by 2 to get the time it took to reach the bottom.

5.0 s/2 = 2.5 s

③ Replace the terms with the values and multiply.

d = (1,530 m/s)(2.5 s)
= 3,825 m

Practice

A sonar signal traveling at 1,490 m/s returns to the ship in 4.0 s. What is the distance to the target?

Review
- Math Practice
- Personal Tutor

Lesson 3
EXPLAIN

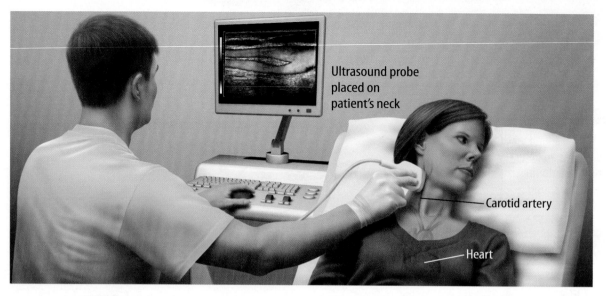

▲ Figure 23 🔑 This ultrasound scanner produces an image of the patient's artery.

REVIEW VOCABULARY

artery
a vessel that takes blood away from the heart

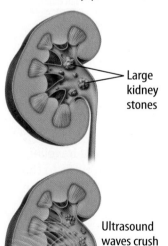

Figure 24 🔑 Ultrasound can break large kidney stones into tiny pieces. ▼

Medical Uses of Ultrasound

Suppose a doctor could see your heart beating or watch blood flowing through your arteries. He or she could determine whether something was wrong without performing surgery. Believe it or not, as **Figure 23** shows, these things are possible using ultrasound.

Ultrasound Imaging Ultrasound scanners work much like sonar. A doctor moves the scanner over different parts of the body. The scanner emits safe, high-frequency sound waves. Body structures such as muscle, fat, blood, and bone reflect sound at different rates. Based on the reflected waves, the scanner produces an image called a sonogram. It is even possible to analyze motion with ultrasound. The scanner in **Figure 23** uses the Doppler effect. By determining how much the frequency of the reflected wave changes, the scanner can determine the blood's speed and direction. Doctors often use sonograms to check the health of unborn babies. Ultrasound is safer than X-rays because it doesn't damage cells.

Treating Medical Problems Has anyone ever massaged your neck or back when it was sore? It not only feels good, but it also relaxes stiff muscles. Many physical therapists use ultrasound to treat joint and muscle sprains or to ease muscle spasms. The vibrations travel through the skin and soft tissue and act like hundreds of tiny fingers massaging the area. Short pulses of high-frequency sound waves can even break apart kidney stones, as shown in **Figure 24**.

🔑 **Key Concept Check** What are some ways in which people use ultrasound?

Lesson 3 Review

 Assessment Online Quiz

Visual Summary

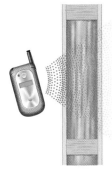

A medium might transmit, absorb, or reflect sound waves that strike it.

Sound waves reflect off hard surfaces in an enclosed space.

Acoustical engineers develop ways to control sound.

FOLDABLES

Use your lesson Foldable to review the lesson. Save your Foldable for the project at the end of the chapter.

What do you think NOW?

You first read the statements below at the beginning of the chapter.

5. Animals use sounds that humans cannot hear to locate objects around them.

6. The method that ships use to locate underwater objects can be used to locate organs or tumors inside the human body.

Did you change your mind about whether you agree or disagree with the statements? Rewrite any false statements to make them true.

Use Vocabulary

① **Describe** reverberation in your own words.

Understand Key Concepts

② Which is the use of sound to identify the distances and positions of objects?
A. echolocation   C. reverberation
B. reflection   D. timbre

③ **Suggest** a way to prevent sounds from echoing in a large gymnasium.

④ **Describe** several ways to limit sound transmission through walls.

Interpret Graphics

⑤ **Identify** conditions that would enable you to hear an echo from the reflected sound waves in the diagram below.

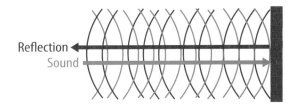

⑥ **Summarize** Copy and fill in a graphic organizer like the one below. Use it to identify and briefly explain three ways sound can interact with matter.

Critical Thinking

⑦ **Assess** the value of echolocation to animals such as bats and dolphins.

Math Skills

Review — Math Practice

⑧ A ship's sonar signal travels at 1,500 m/s and returns 0.40 s after it is sent. How deep is the shipwreck the ship locates?

Lesson 3 • 589
EVALUATE

Inquiry Lab

90 minutes

Make Your Own Musical Instrument

Materials

ruler

rubber bands

scissors

balloons

drinking straws

string

Also needed:
tape, pencil, bottles, boxes, plastic wrap, aluminum pans, paper, metal cans, tubing, cardboard tubes

Safety

Suppose you are a professional musician and want to design a new instrument. What type of instrument will you make? Your instrument must be able to produce a range of pitches. It also must be able to play loudly or quietly. The instrument must be durable enough to play over and over. In this lab, you will make a model of your instrument. Then you will modify it, one part at a time, until you are satisfied with the result.

Question
What factors should you consider when making a musical instrument?

Procedure
1. Read and complete a lab safety form.
2. Design a musical instrument made from common materials. Your instrument must be able to play a range of frequencies and a range of amplitudes. Record your design in your Science Journal.
3. Have your teacher approve your design before you build your instrument.
4. Build your instrument, and then play it.
5. Think about how you could improve the design of your instrument. Record your observations and ideas in your Science Journal.

6. Make a modification to your instrument. Remember that in any investigation you should change only one variable at a time—the independent variable. In your Science Journal, record what you changed and why you changed it.

7. Play the instrument. Did your modification produce the change in sound you intended? Why or why not? Record your observations.

8. Continue to improve your instrument as time permits. You may change the same variable or a different one, but be sure to change only one variable at a time. Record your observations and details about all changes.

Analyze and Conclude

9. **Classify** your instrument as a string instrument, a wind instrument, or a percussion instrument. Identify the properties you used to classify the instrument.

10. **Draw** a sketch of your instrument. Label the parts of the instrument that vibrate to produce sounds.

11. **Compare** your instrument to a similar common musical instrument.

12. **Critique** your design, your changes, and the results.

13. **The Big Idea** Describe how your instrument produces sounds.

Communicate Your Results

Present your musical instrument to the class, and explain your design. Demonstrate your instrument by playing a simple song.

Common musical instruments are both durable and able to be tuned. Explore what it means to tune an instrument. In what ways can you make your instrument playable by a professional musician?

Lab Tips

☑ Think about how each type of instrument changes the frequency and amplitude of sound.

☑ When you decide to change something about your instrument, consider whether the change might affect frequency, amplitude, and durability.

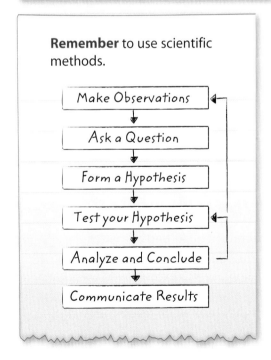

Remember to use scientific methods.

- Make Observations
- Ask a Question
- Form a Hypothesis
- Test your Hypothesis
- Analyze and Conclude
- Communicate Results

Chapter 16 Study Guide

THE BIG IDEA — Vibrations in matter produce sound waves. Each sound source creates a unique wave form that the ear can identify. People use the way sound waves are absorbed, transmitted, and reflected by matter in many ways.

Key Concepts Summary | Vocabulary

Lesson 1: Producing and Detecting Sound

- **Vibrations** in a medium produce **sound waves.**
- Sound waves travel as **compressions** and **rarefactions.**
- Sound waves travel faster through materials in which the particles are closer together.
- The outer ear collects sound. The middle ear amplifies sound. The inner ear converts vibrations to nerve signals.

sound wave p. 565
longitudinal wave p. 565
vibration p. 565
medium p. 566
compression p. 566
rarefaction p. 566

Lesson 2: Properties of Sound Waves

- The greater a sound wave's energy, the larger the **amplitude** and the greater the wave's **intensity.**
- Sound waves with a longer **wavelength** have a lower **frequency** and a lower **pitch.**
- Different frequencies of sound waves combine and form a complex wave the brain recognizes.
- Strings, air columns, or surfaces of instruments vibrate and produce music.

amplitude p. 573
intensity p. 574
wavelength p. 575
frequency p. 575
pitch p. 575
Doppler effect p. 576
interference p. 576
resonance p. 578

Lesson 3: Using Sound Waves

- Sound waves can be transmitted, **reflected,** or **absorbed** by matter.
- Materials and shapes in a room can improve vibrations, absorb excess sound waves, and reflect sound waves to fill the room.
- Ultrasound is used for medical imaging and treatment.

absorption p. 584
reflection p. 584
echo p. 584
reverberation p. 585
acoustics p. 585
echolocation p. 587
sonar p. 587

Study Guide

- Personal Tutor
- Vocabulary eGames
- Vocabulary eFlashcards

Chapter Project

Assemble your lesson Foldables as shown to make a Chapter Project. Use the project to review what you have learned in this chapter.

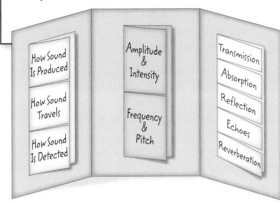

Use Vocabulary

1. Describe a longitudinal wave.

2. The part of a sound wave in which the particles are most spread out is called a(n) _____.

3. Explain the relationship between frequency and pitch.

4. A change in frequency called the _____ depends on the motion of the sound source and the position of the listener.

5. Describe how resonance works.

6. The collection of sound reflections in a closed space is called _____.

7. An example of _____ is when a dolphin emits loud clicks and listens for the echo.

Link Vocabulary and Key Concepts

Concepts in Motion — Interactive Concept Map

Copy this concept map, and then use vocabulary terms from the previous page to complete the concept map.

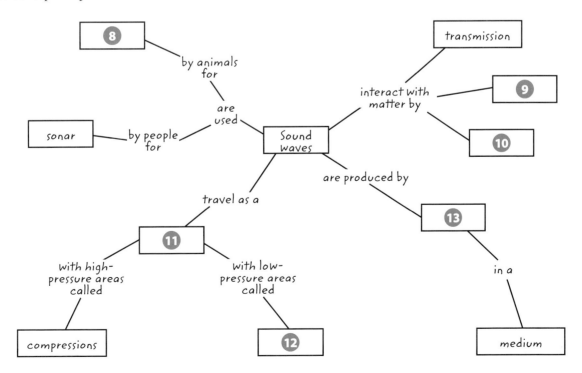

Chapter 16 Study Guide • 593

Chapter 16 Review

Understand Key Concepts

1. Which type of matter would transmit sound waves fastest?
  A. air at 5°C
  B. air at 20°C
  C. ice at 0°C
  D. seawater at 20°C

2. A large windowpane on a storefront vibrates as a large truck drives by. Which type of interaction between sound and matter best explains this vibration?
  A. absorption
  B. resonance
  C. reverberation
  D. transmission

3. As _____ decreases, sound intensity decreases.
  A. amplitude
  B. quality
  C. wave speed
  D. wavelength

4. Wave frequency is measured in which unit?
  A. decibel
  B. hertz
  C. meter
  D. second

5. What causes the sound waves's intensity to decrease in the picture below?

  A. insulation
  B. interference
  C. rarefaction
  D. resonance

6. Which property describes the distance between two identical points on a sound wave?
  A. amplitude
  B. frequency
  C. pitch
  D. wavelength

7. When you hear a sound, its pitch is mostly influenced by which properties of a sound wave?
  A. amplitude and speed
  B. frequency and amplitude
  C. speed and frequency
  D. wavelength and frequency

8. Which shows overtones of a fundamental?

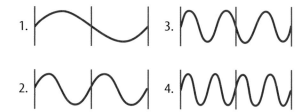

  A. only 1
  B. 1 and 2
  C. 2, 3, and 4
  D. 1, 2, 3, and 4

9. Which property of sound waves changes when you increase the volume on a car radio?
  A. amplitude
  B. frequency
  C. speed
  D. wavelength

10. Which is the region of a sound wave where particles of a medium are most spread out?
  A. compression
  B. rarefaction
  C. reverberation
  D. transmission

Chapter Review

Assessment — Online Test Practice

Critical Thinking

11 **Visualize** A bat in a dark cave sends out a high-frequency sound wave. The bat detects an increase in frequency after the sound bounces off its prey. Describe the possible motion of the bat and its prey.

12 **Synthesize** An MP3 player at maximum volume produces sound at 110 dB. The table below shows the recommended time exposure before risk of damage to the ear. How many hours a day could you listen to your MP3 player at full volume without risking hearing loss? Explain.

Recommended Noise Exposure Limits	
Sound Level (dB)	Time Permitted (hr)
90	8
95	4
100	2
105	1

13 **Construct** a diagram of two sound waves of equal amplitude and frequency that experience destructive interference.

14 **Create** a diagram that includes drawings of sound waves to show how the ear distinguishes among sounds, such as different voices or musical instruments.

15 **Compare** the effects on the ear of a single very loud sound of 150 dB with prolonged exposure to sounds of 90 dB.

Writing in Science

16 **Write** A friend tells you that his family always complains about the noise when he practices his electric guitar in his room. Write at least four recommendations for how your friend might reduce the sound that escapes from his room.

REVIEW THE BIG IDEA

17 Explain how sound is produced, travels from one place to another, and is detected by the ear.

18 Look at all the equipment in the music control room shown in the photograph. Think about what you have learned about sound in the chapter. How can musicians produce, describe, and use sound?

Math Skills

Review — Math Practice

Use a Formula

19 A ship sends out a sonar signal to locate other ships. Traveling at 1,530 m/s, a signal returns to the ship 3.6 s after it is sent. What is the distance from the other ship?

20 A sonar signal takes 3 s to return from a sunken ship directly below. Find the depth of the sunken ship if the speed of the signal is 1,440 m/s.

21 A ship is sailing in water that is 500 m deep. Sound travels at 1,520 m/s in the body of water. A sonar echo returns to the ship in 0.6 s. Did the signal bounce off the bottom? Explain.

Standardized Test Practice

Record your answers on the answer sheet provided by your teacher or on a sheet of paper.

Multiple Choice

Use the figure below to answer questions 1–3.

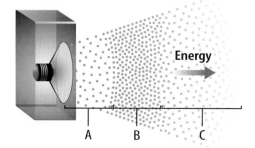

1 How does this speaker cause sound that can be heard?

- **A** It produces echoes.
- **B** It produces vibrations.
- **C** It releases light.
- **D** It releases heat.

2 What type of wave does the speaker produce for sound to be heard?

- **A** electromagnetic
- **B** longitudinal
- **C** surface
- **D** transverse

3 Point A in the figure is in a rarefaction. Point B is in a compression. Point C is a location in the air at normal air pressure before the sound wave has arrived. Which statement is true of the pressure at points A, B, and C?

- **A** The pressure at point A is the greatest.
- **B** The pressure at point B is the greatest.
- **C** The pressure at point C is the greatest.
- **D** The pressure at all three points is equal.

4 Which two factors can affect the speed of sound?

- **A** size and density of the medium
- **B** mass and density of the medium
- **C** temperature and mass of the medium
- **D** temperature and density of the medium

Use the table below to answer question 5.

Speed of Sound in Different Materials	
Material (at 20°C)	Speed (m/s)
Air	343
Glass	5,640
Iron	5,130
Water	1,481

5 A sound wave takes about 0.03 s to move through a material that is 10.3 m long. What is the material?

- **A** air
- **B** glass
- **C** iron
- **D** water

6 Which structure in the human ear transfers sound vibrations to the stirrup?

- **A** anvil
- **B** cochlea
- **C** ear canal
- **D** eardrum

7 Which type of sound wave has the greatest energy?

- **A** a wave that has a very high intensity
- **B** a wave that has a very low amplitude
- **C** a wave with particles that do not vibrate from their rest positions
- **D** a wave with particles that vibrate a very short distance from their rest positions

Standardized Test Practice

Use the figure below to answer questions 8 and 9.

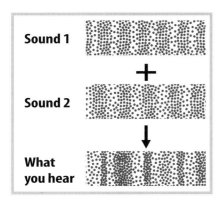

8. What can you tell about sounds 1 and 2 by looking at the figure?
   A Sound 1 has a lower pitch.
   B Sound 1 has a greater frequency.
   C Sound 1 has a smaller amplitude.
   D Sound 1 has a greater wavelength.

9. What does the bottom wave in the figure represent?
   A silence
   B interference
   C the Doppler effect
   D a fundamental frequency

10. What differs when the same note is played with the same amplitude by two different kinds of instruments?
    A energy
    B fundamentals
    C intensity
    D overtones

Constructed Response

Use the figure below to answer questions 11 and 12.

11. Describe the type of sound technology being used in the figure. How does this type of technology differ from echolocation?

12. The ship in this figure sends out a signal that returns to the ship in 4 s. If sound travels through seawater at 1,530 m/s, how far away from the ship are the fish?

13. Why would a sound studio have soft foam squares on the walls?

14. Name two ways that ultrasound is used to detect medical problems. Name two ways that ultrasound is used to treat medical problems.

NEED EXTRA HELP?													
If You Missed Question...	1	2	3	4	5	6	7	8	9	10	11	12	13	14
Go to Lesson...	1	1	1	1	1	1	2	2	2	2	3	3	3	3

Chapter 17

Electromagnetic Waves

 How can you describe and use electromagnetic waves?

Inquiry Invisible Waves?

The Sun constantly emits energy as electromagnetic waves. However, the human eye cannot see some of these waves. A computer was used to add color to the images of invisible solar waves shown here. Scientists use similar images to determine the temperature of objects. In this chapter, you will learn about electromagnetic waves and that all objects, even you, emit this form of energy.

- Why do you think all objects give off electromagnetic waves?
- How do photographs such as these help scientists study the Sun?
- How can you describe and use electromagnetic waves?

Get Ready to Read

What do you think?
Before you read, decide if you agree or disagree with each of these statements. As you read this chapter, see if you change your mind about any of the statements.

1. Warm objects emit radiation, but cool objects do not.
2. Light always travels at a speed of 300,000 km/s.
3. Red light has the least amount of energy of all colors of light.
4. A television remote control emits radiation.
5. Thermal images show differences in the amount of energy people or objects give off.
6. When you call a friend on a cell phone, a signal travels directly from your phone to your friend's phone.

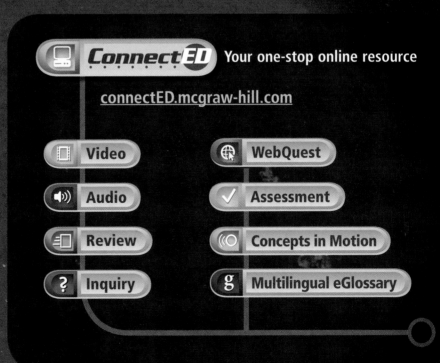

Lesson 1

Electromagnetic Radiation

Reading Guide

Key Concepts 🗝
ESSENTIAL QUESTIONS
- How do electromagnetic waves form?
- What are some properties of electromagnetic waves?

Vocabulary
electromagnetic wave p. 601
radiant energy p. 601

 Multilingual eGlossary

 Video

What's Science Got to do With It?

Inquiry Catching Waves?

Camping in a remote area far from electric lines can be a challenge. Is it possible to get energy for things such as lights and heating? This camper can use these solar panels to capture energy from the Sun. Because electromagnetic waves can travel through space, the Sun is Earth's most important source of energy, both in cities and in remote areas.

Inquiry Launch Lab

15 minutes

How can you detect invisible waves?

You can see light from the Sun, but other forms of the Sun's energy are invisible. One way to detect invisible waves is to observe their effects on things.

1. Read and complete a lab safety form.
2. With a **marker,** label a **clear plastic cup** *TAP* near the bottom of the cup. Fill it with tap water.
3. Label another cup *TONIC,* and fill it with **tonic water.**
4. Hold the cup with tap water near a **lamp.** Hold a sheet of **black construction paper** behind the cup. Observe the water as you slowly move the cup and paper away from and then closer to the lamp several times. Do you notice any change? Record observations in your Science Journal.
5. Repeat step 4 with the tonic water. Record your observations.
6. Repeat steps 4 and 5, but this time move each cup and paper closer to and then away from bright sunlight instead of a lamp. Record your observations.

Think About This

1. How was the effect of sunlight different from the effect of lamplight on each type of water?
2. **Key Concept** How did your results show that sunlight emits invisible waves?

What are electromagnetic waves?

Suppose you live in a remote location. Would you be able to have electric lights? Could you watch television? Thanks to the Sun, the answer is yes! Like the camper on the previous page, you could use solar panels to capture the Sun's energy and transform it into electricity.

Energy from the Sun reaches Earth by traveling in waves. Many types of waves can travel only through a medium, or matter. Waves in a pond, for example, require water to travel. Waves from the Sun are different because they can travel through empty space. *A wave that can travel through empty space and through matter is called an* **electromagnetic wave.** These waves radiate, or spread out, in all directions from a source. *Energy carried by an electromagnetic wave is called* **radiant energy.** This energy also is known as electromagnetic radiation.

You will read in this lesson that the Sun is not the only source of radiant energy. However, the Sun is the source that provides most of Earth's radiant energy. You also will read about how electromagnetic waves form, and learn about some properties of electromagnetic waves.

Reading Check What is electromagnetic radiation?

FOLDABLES

Create a vertical three-tab book. Label it as shown. Use it to organize your notes about electromagnetic waves.

WORD ORIGIN

radiant
from Latin *radiantem*, means "shining"

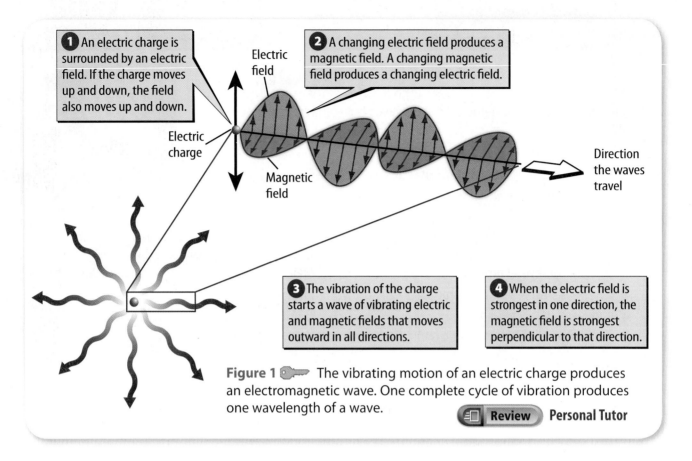

Figure 1 The vibrating motion of an electric charge produces an electromagnetic wave. One complete cycle of vibration produces one wavelength of a wave.

How Electromagnetic Waves Form

Electromagnetic waves form when an electric charge accelerates by either speeding up, slowing down, or changing direction. This happens when a charged particle vibrates, as shown in **Figure 1**.

Force Fields If you ever have played with a magnet, you know that it is surrounded by a field, or area, where the force of the magnet is present. The same is true for a charged particle, such as an electron. An electric field surrounds the charged particle.

Connected Fields Scientists have found that electric fields and magnetic fields are related.

- A changing electric field produces a magnetic field.
- A changing magnetic field produces a changing electric field.

As a charged particle vibrates, the electric field around it vibrates. This changing electric field produces a magnetic field. As the magnetic field changes, it produces a changing electric field. These connected fields spread out in all directions as electromagnetic waves. Just as shaking the end of the rope in **Figure 2** produces a wave, a vibrating charge produces a wave.

Key Concept Check How do electromagnetic waves form?

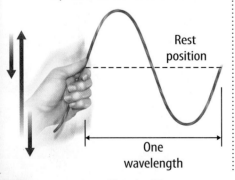

Figure 2 Shaking the rope up and down one time produces one wavelength. A charged particle also produces one wavelength when it moves up and down one time. ▼

Properties of Electromagnetic Waves

Electromagnetic waves usually are drawn as a single curve, like the rope in **Figure 2**. The waves are called transverse because the disturbance is perpendicular to the direction they travel.

Wavelength and Frequency As with all waves, the wavelength and frequency of electromagnetic waves are related. As shown in **Figure 2**, wavelength is the distance between one point on a wave to the nearest point just like it. Frequency is the number of wavelengths that pass by a point in a certain period of time, such as a second. As frequency decreases, wavelength increases. You will read in Lesson 2 that electromagnetic waves are grouped according to wavelength and frequency.

Wave Speed Electromagnetic waves travel through space at 300,000 km/s, or the speed of light (*c*). A wave's speed (*s*) is its frequency (*f*) multiplied by its wavelength (λ). To determine the wavelength of a wave moving through space, divide the speed of light (*c*) by the frequency of the wave, as shown in the Math Skills box.

What happens when electromagnetic waves move through matter? Suppose you run across a beach toward a lake or an ocean. You move quickly over the sand, but you slow down in the water. Electromagnetic waves behave similarly. When they encounter matter, electromagnetic waves slow down.

 Key Concept Check What are some properties of electromagnetic waves?

Math Skills

Solve One-Step Equations

Use inverse operations to keep sides of an equation equal. What is the wavelength of an electromagnetic wave in space that has a frequency of 500,000 Hz?

a. Use the wave-speed equation: **wave speed = frequency × wavelength**:

$$s = f\lambda$$

b. Divide by frequency. Then substitute known values. (Hint: Hz = 1/s)

$$\lambda = \frac{s}{f} = \frac{300{,}000 \text{ km/s}}{500{,}000 \text{ Hz}} = 0.6 \text{ km}$$

Practice
What is the wavelength of an electromagnetic wave with a frequency of 125,000 Hz?

- Math Practice
- Personal Tutor

Inquiry MiniLab — 10 minutes

How are electric fields and magnetic fields related?

A compass can show the effect of an electric field on a magnetic field.

1. Read and complete a lab safety form.
2. Place a **battery** on a table, as shown in the photo. Place a **compass** a few centimeters from one end of the battery.
3. Tap the ends of a **20-cm wire** to each end of the battery and quickly remove them. What happens to the compass? Record your observations in your Science Journal.

   ⚠ Do not leave the wire attached to the battery. It can get very hot.

4. Repeat step 3 several times. Then hold the wire to the battery for only 2 seconds. Record your observations of the compass.

Analyze and Conclude

1. **Contrast** the effect on the compass when you touched the wire quickly and slowly.
2. **Key Concept** How do your results relate electric fields and magnetic fields?

Figure 3 Most of the Sun's energy is carried by infrared waves. ▶

Visual Check What percentage of the Sun's electromagnetic waves are either light or infrared waves?

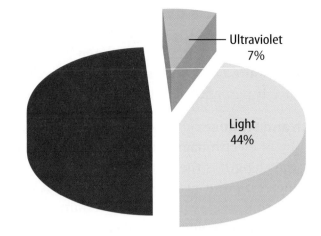

Ultraviolet 7%
Light 44%

REVIEW VOCABULARY

temperature
a measure of the average kinetic energy of the particles that make up an object

Figure 4 A supernova is the result of an exploding star. Colors were added with a computer to show the different wavelengths of electromagnetic waves emitted by this supernova. ▼

Sources of Electromagnetic Waves

When you hear the term *electromagnetic radiation, or radiant energy,* you might imagine dangerous rays that you should avoid. It might surprise you to learn, however, that you are a source of electromagnetic waves! All matter contains charged particles that constantly vibrate. As a result, all matter—including you—produces electromagnetic waves, and therefore radiant energy. As an object's temperature increases, its particles vibrate faster. Therefore, the object produces electromagnetic waves with a greater frequency, or shorter wavelengths.

The Sun

Earth's most important energy source is the Sun. The Sun emits energy by giving off electromagnetic waves. Only a tiny amount of these waves reach Earth. The Sun has many areas with different temperatures and produces electromagnetic waves with many different wavelengths. As shown in **Figure 3**, almost all of the Sun's energy is carried by three types of waves—ultraviolet waves, light waves, and infrared waves. Light waves are the only type of electromagnetic waves you can see. Infrared waves have wavelengths longer than light. Ultraviolet waves have wavelengths shorter than light. You will learn more about the different types of electromagnetic waves in Lesson 2.

Other Sources of Electromagnetic Waves

Even though the Sun is Earth's most important source of electromagnetic radiation, it is not the only one. All matter, both in space and on Earth, produces electromagnetic waves.

Sources in Space If you look up at the night sky, you see the Moon, stars, and planets. Telescopes on Earth and on satellites above Earth's atmosphere produce images of radiation emitted by these objects. Some of the radiation is visible, but most of it is not. **Figure 4** shows what radiation emitted by an exploding star might look like if your eyes could detect it.

Sources on Earth What do a campfire, a lightbulb, and a burner on an electric stove have in common? They all are hot enough to produce electromagnetic waves that carry energy you can feel. Look around you right now. Your book, the wall, people, and everything else you see produce electromagnetic waves, too. Some waves you detect, but others you do not. Telescopes produce visible images of radiation from space. Also, special cameras produce visible images of invisible waves on Earth. As shown in **Figure 5,** the ultraviolet waves from a flower produce an image very different from the waves you see with your eyes.

The Energy of Electromagnetic Waves

Have you ever seen someone with a terrible sunburn, such as the boy in **Figure 6?** Some of the Sun's waves have enough energy to damage your skin. The energy of electromagnetic waves is related to their frequency. Waves that have a higher frequency, such as ultraviolet waves, have higher energy. Waves that have a lower frequency, such as light waves and infrared waves, have lower energy. Light from the girl's flashlight in **Figure 6** can never damage her skin because the light waves do not have enough energy.

The relationship between energy and other wave properties is different for mechanical waves and electromagnetic waves. The energy of a mechanical wave is related to its amplitude. A water wave, for example, with a high amplitude has a lot of energy. The energy of electromagnetic waves is related to their frequency, not amplitude. As the frequency of an electromagnetic wave increases, the energy of the wave increases.

 Reading Check Why can the Sun's rays cause a burn, but the light from a flashlight cannot harm your skin?

▲ **Figure 5** The image on the bottom shows what the dandelions would look like if your eyes could detect the ultraviolet waves the flowers emit.

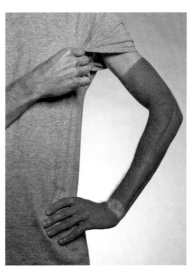

◀ **Figure 6** Ultraviolet waves from the Sun carry enough energy to damage skin cells. Light waves from a flashlight do not carry enough energy to cause damage.

Lesson 1 Review

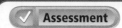 Assessment   Online Quiz

Visual Summary

An accelerating charge produces an electromagnetic wave similar to the wave produced by shaking the end of a rope.

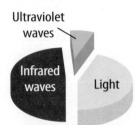

Almost all of the Sun's energy travels by infrared waves, light waves, or ultraviolet waves.

Electromagnetic waves transfer energy from one place to another, even through empty space.

FOLDABLES

Use your lesson Foldable to review the lesson. Save your Foldable for the project at the end of the chapter.

What do you think NOW?

You first read the statements below at the beginning of the chapter.

1. Warm objects emit radiation, but cool objects do not.
2. Light always travels at a speed of 300,000 km/s.

Did you change your mind about whether you agree or disagree with the statements? Rewrite any false statements to make them true.

Use Vocabulary

1. **Use the term** *radiant energy* in a sentence.

2. Electromagnetic waves carry _____.

Understand Key Concepts

3. What is the speed of electromagnetic waves in space?
   A. 30,000 m/s   C. 300,000 m/s
   B. 30,000 km/s   D. 300,000 km/s

4. **Identify** What must happen in order for an electromagnetic wave to form?

Interpret Graphics

5. **Examine** the electromagnetic wave below.

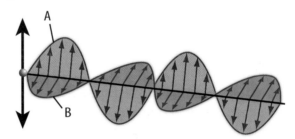

Identify regions A and B in the figure. Explain what causes these regions.

6. **Sequence** Copy and fill in a graphic organizer like the one below to describe how an electromagnetic wave forms and travels away from a source.

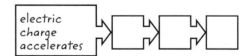

Critical Thinking

7. **Create** a poster that explains how electromagnetic waves sometimes behave like a stream of particles.

Math Skills

— Math Practice —

8. What is the wavelength of an electromagnetic wave with a frequency of 135,000 Hz?

Solar Sails
Using Light to Sail Through Space

How far and how fast can a spacecraft travel? Scientists are studying an exciting way to travel beyond the solar system. A technology called solar sails would allow a spacecraft to travel faster and farther than ever before possible.

Johannes Kepler first proposed solar sails about 400 years ago. He observed comets' tails facing away from the Sun and concluded that sunlight caused pressure—a force that might be harnessed with sails.

The Power of the Sun

Solar sails use light, not matter, to propel spacecraft. When reflected off a solar sail, light transfers its momentum to the surface of the sail, giving it a slight push. The solar sail moves a spacecraft very slowly at first. However, as long as there is a steady stream of light, the sail can eventually accelerate a spacecraft to incredibly high speeds.

The future of solar sail technology is unknown. So far, the use of solar sails has been successful only in a laboratory test chamber. The image below shows how scientists believe a solar sail could work if used in space.

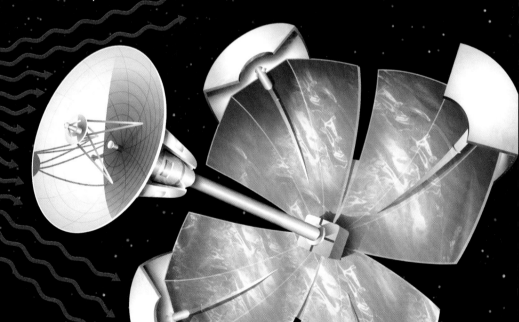

Characteristics of a Solar Sail

- Large: at least as big as a football field
- Lightweight: 40–100 times thinner than a sheet of paper
- Very shiny: push provided by reflected light
- Rigid and durable: lasts many years in deep space

It's Your Turn

CALCULATE After traveling continuously for 3 years, a solar sail could reach a speed of 160,000 km/h. The speed of light is about 300,000 km/s. First calculate the kilometers per second a solar sail could reach. Then compare that number to the speed of light. What comparison can you make?

Lesson 2

The Electromagnetic Spectrum

Reading Guide

Key Concepts
ESSENTIAL QUESTIONS

- What is the electromagnetic spectrum?
- How do electromagnetic waves differ?

Vocabulary

electromagnetic spectrum p. 609
radio wave p. 610
microwave p. 610
infrared wave p. 610
ultraviolet wave p. 611
X-ray p. 611
gamma ray p. 611

 Multilingual eGlossary

 Video

What's Science Got to do With It?

Inquiry) How Many Colors?

Can you name the colors of the rainbow? You might know the familiar colors from red to violet, but no one can name all the colors. Rainbows are a continuous range of colors. Each color is an electromagnetic wave with a slightly different wavelength. The range of electromagnetic waves extends to longer and shorter wavelengths you cannot see. What are other types of electromagnetic waves?

Inquiry Launch Lab

20 minutes

How do electromagnetic waves differ?

Sunlight contains many types of invisible electromagnetic waves. How are these waves different from light you can see?

1. Read and complete a lab safety form.
2. Obtain **four beads.** Place each in a separate **small, self-sealing plastic bag.**
3. Obtain **three types of sunscreen.** Each type should have a different sun protection factor (SPF).
4. Use a **permanent marker** to write the SPF rating on one side of each bag. On the other side, apply a thin layer of sunscreen that has that SPF. Label the fourth bag *No Sunscreen*.
5. Place the bags and beads near a **lamp** with the sunscreen side up. Observe the beads for several minutes. Record your observations in your Science Journal.
6. Place the bags outside in sunlight. Observe the beads again for several minutes, and record your observations.

Think About This

1. Contrast what happened to the beads when you placed them near lamplight and in sunlight.
2. Describe the relationship between the SPF numbers and what happened to the beads.
3. **Key Concept** What do you think could have caused your results?

What is the electromagnetic spectrum?

The waves that carry voices to your cell phone, the waves of energy that toast your bread, and the X-rays that a dentist uses to check the health of your teeth are all electromagnetic waves. The changing motion of an electric charge produces each type of electromagnetic wave. Each type of wave has a different frequency and wavelength, and each carries a different amount of energy. Electromagnetic waves might vibrate from a thousand times a second to trillions of times a second. They might be as large as a house or as small as an atom's nucleus. *The* **electromagnetic spectrum** *is the entire range of electromagnetic waves with different frequencies and wavelengths.*

Key Concept Check What is the electromagnetic spectrum, and how do electromagnetic waves differ?

Classifying Electromagnetic Waves

There are many shades among the familiar colors of a rainbow. Each color gradually becomes another. However, the electromagnetic spectrum is organized into groups based on the wavelengths and frequencies of the waves. Like the colors of a rainbow, each group blends into the next. You will read about how electromagnetic waves are grouped on the next pages.

WORD ORIGIN

spectrum
from Latin *spectrum*, means "appearance"

FOLDABLES

Use four sheets of paper to make an eight-layer book. Label it as shown. Use it to compare and contrast the different types of electromagnetic waves.

Lesson 2
EXPLORE

Electromagnetic Spectrum

A **radio wave** *is a low-frequency, low-energy electromagnetic wave that has a wavelength longer than about 30 cm.* Some radio waves have wavelengths as long as a kilometer or more. Radio waves often are used for communication. The wavelengths are long enough to move around many objects, but the energy is low enough that they aren't harmful. On Earth, radio waves usually are produced by an electric charge moving in an antenna, but the Sun and other objects in space also produce radio waves.

1. Radio waves   2. Microwaves   3. Infrared waves

Longer wavelengths, Lower frequencies, Lower energy

A **microwave** *is a low-frequency, low-energy electromagnetic wave that has a wavelength between about 1 mm and 30 cm.* Like radio waves, microwaves are used for communication, such as cell phone signals. With shorter wavelengths than radio waves, microwaves are less often scattered by particles in the air. Microwaves are useful for satellite communications because they can pass through Earth's upper atmosphere. Because of the frequency range of microwaves, food molecules such as water and sugar can absorb their energy. This makes microwaves useful for cooking.

An **infrared wave** *is an electromagnetic wave that has a wavelength shorter than a microwave but longer than light.* Vibrating molecules in any matter emit infrared waves. Even your body emits infrared radiation. You cannot see infrared waves, but if you warm your hands near a campfire, you can feel them. Your skin senses infrared waves with longer wavelengths as warmth. Infrared waves with shorter wavelengths do not feel warm. Your television remote control, for example, sends out these waves.

Figure 7 Electromagnetic waves have a wide range of uses because of their different wavelengths, frequencies, and energy.

Visual Check Which types of electromagnetic waves have wavelengths too long for your eyes to see?

④ Light is electromagnetic waves that your eyes can see. You might describe light as red, orange, yellow, green, blue, indigo, and violet. Red light has the longest wavelength and the lowest frequency. Violet light has the shortest wavelength and the highest frequency. Each name represents a family of colors, each with a range of wavelengths.

④ **Light waves**　⑤ **Ultraviolet waves**　⑥ **X-rays**　⑦ **Gamma rays**

Shorter wavelengths, Higher frequencies, Higher energy

⑤ An **ultraviolet wave** *is an electromagnetic wave that has a slightly shorter wavelength and higher frequency than light and carries enough energy to cause chemical reactions.* Earth's atmosphere prevents most of the Sun's ultraviolet rays from reaching Earth. But did you know that you can get a sunburn on a cloudy day? This is because ultraviolet waves carry enough energy to move through clouds and to penetrate the skin. They can damage or kill cells, causing sunburn or even skin cancer.

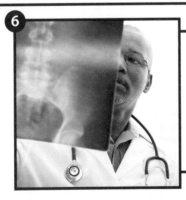

⑥ An **X-ray** *is a high-energy electromagnetic wave that has a slightly shorter wavelength and higher frequency than an ultraviolet wave.* Have you ever had an X-ray taken to see if you had a broken bone? X-rays have enough energy to pass through skin and muscle, but the calcium in bone can stop them. Scientists learn about objects and events in space, such as black holes and star explosions, by studying the X-rays they emit.

⑦ A **gamma ray** *is a high-energy electromagnetic wave with a shorter wavelength and higher frequency than all other types of electromagnetic waves.* Gamma rays are produced when the nucleus of an atom breaks apart or changes. As shown in the photo, gamma rays have enough energy that physicians can use them to destroy cancerous cells. Like X-rays, gamma rays form in space during violent events, such as the explosion of stars.

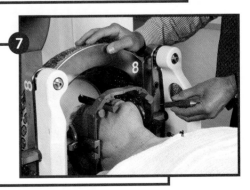

Lesson 2
EXPLAIN

Lesson 2 Review

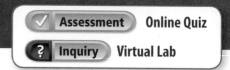

Visual Summary

Light represents only a small portion of the total electromagnetic spectrum.

Because of differences in wavelength, frequency, and energy, electromagnetic waves have many common uses.

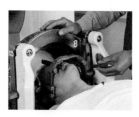

High-frequency electromagnetic waves carry so much energy that they often are used for medical imaging and treatment.

FOLDABLES

Use your lesson Foldable to review the lesson. Save your Foldable for the project at the end of the chapter.

What do you think NOW?

You first read the statements below at the beginning of the chapter.

3. Red light has the least amount of energy of all colors of light.

4. A television remote control emits radiation.

Did you change your mind about whether you agree or disagree with the statements? Rewrite any false statements to make them true.

Use Vocabulary

1. Explain the difference between a microwave and a radio wave.

2. Explain the difference between an infrared wave and an ultraviolet wave.

Understand Key Concepts

3. Which electromagnetic waves have the highest energy?
- A. gamma rays
- B. light waves
- C. radio waves
- D. X-rays

4. Name the types of waves that make up the electromagnetic spectrum, from longest wavelength to shortest wavelength.

5. Compare the frequency of gamma waves with the frequency of light waves.

Interpret Graphics

6. Identify The diagram below shows three types of electromagnetic waves approaching a human body.

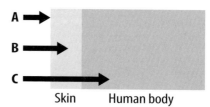

Which represents X-rays? Which represents ultraviolet waves? Explain.

7. Sequence Draw a graphic organizer like the one below. In the boxes, write the types of electromagnetic waves from lowest to highest energy. Add boxes as necessary.

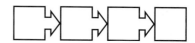

Critical Thinking

8. Compare and contrast infrared waves, light, and ultraviolet waves.

9. Infer Why are X-rays and gamma rays able to penetrate the body, but radio waves are not?

612 Chapter 17
EVALUATE

Inquiry Skill Practice: Test a Hypothesis

30 minutes

What's at the edge of a rainbow?

Materials

- prism
- box
- thermometers

Safety

If you have been to a fast food restaurant or cafeteria, you might have noticed special lights hanging over the french fries to keep them warm. Why are the lightbulbs usually red?

Learn It

A hypothesis is a possible explanation for an observation. It might be based on things you have seen, as well as knowledge you already have. You **test a hypothesis** with an experiment. You make more observations and collect data. Your data results either support or do not support your hypothesis.

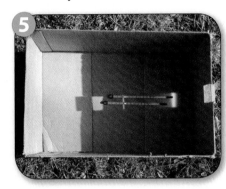

Try It

1. Read and complete a lab safety form.

2. Mount a prism on the end of a large box, as shown in the photo.

3. Adjust the prism and box so that the prism catches the Sun's rays and produces a spectrum on the bottom of the box. You might need to tilt the box.

4. Think about the lights that keep food warm at a restaurant. Based on your observations, form a hypothesis about what part of the electromagnetic spectrum is most useful for keeping food warm and why.

5. Place two thermometers in the box. One should be in the visible part of the spectrum, and the other should be just past the red end of the spectrum.

6. Create a chart like the one below in your Science Journal. Monitor and record the temperatures shown on the thermometers each minute for 5 minutes.

Apply It

7. **Design** a double-line graph to compare the temperature readings of each thermometer.

8. **Conclude** Did your data results support your hypothesis? Why or why not?

9. **Decide** How do you think a thermometer would respond if you placed it just past the violet portion of the spectrum? Explain.

10. **Key Concept** Describe the differences between the types of electromagnetic waves that reach each thermometer.

Minute	Thermometer 1 (in color)	Thermometer 2 (in infrared)
0 min		
1 min		
2 min		
3 min		
4 min		
5 min		
Change in temperature		

Lesson 2 EXTEND

Lesson 3

Reading Guide

Key Concepts 🔑
ESSENTIAL QUESTIONS

- How are different types of electromagnetic waves used for communication?
- What are some everyday applications of electromagnetic waves?
- What are some medical uses of electromagnetic waves?

Vocabulary
broadcasting p. 615
carrier wave p. 616
amplitude modulation p. 616
frequency modulation p. 616
Global Positioning System p. 618

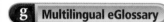

Multilingual eGlossary

Video BrainPOP®

Using Electromagnetic Waves

Inquiry Glowing Fingerprint?

A glowing fingerprint might look strange, but to a crime-scene detective, it's a clue. Detectives dust the scene with a powder that glows when ultraviolet waves strike it. If the powder sticks to a handprint, using ultraviolet waves might solve the crime. What are some other uses of electromagnetic waves?

Inquiry Launch Lab

15 minutes

How do X-rays see inside your teeth?

Dentists take X-rays of your mouth to see if you have any problems. How does an X-ray produce images of hard materials such as teeth?

1. Read and complete a lab safety form.
2. Use **scissors** to cut an **index card** into four teeth shapes. Put the paper teeth in different places on a **piece of black paper** on the bottom of a **shoe box**.
3. Cover the shoebox opening with a piece of **screen**. Use **rubber bands** to attach the screen to the top of the shoe box.
4. Sprinkle **flour** evenly across the surface of the screen. Use a **toothbrush** to gently brush the flour on the screen until all of it falls through the screen.
5. Carefully remove the screen and remove the card parts from the box.
6. Observe the paper "X-ray." Record your observations in your Science Journal.

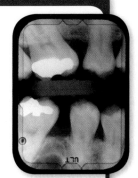

Think About This

1. What did you observe in the bottom of the box?
2. **Key Concept** If the screen represents your skin, muscle, and fatty tissues, what do you think this activity tells you about how X-ray machines form images?

How do you use electromagnetic waves?

There are few things you do that don't involve some type of electromagnetic wave. Think about a typical day. Perhaps a clock radio wakes you in the morning with your favorite music. You are warm under your blanket because infrared radiation from your body transforms to thermal energy that is trapped under your blanket. When you finally crawl out of bed, you turn on a lamp and use light to see which clothes to wear. These are just a few of the ways you use electromagnetic waves. In this lesson, you will read about common uses for all types of electromagnetic waves.

Radio Waves

It probably is no surprise to you that radio and television stations use radio waves to send out signals. Radio waves can pass through many buildings, yet they don't carry enough energy to harm humans. The wavelengths of radio waves are long enough to go around many obstacles. They travel through air at the same speed as all electromagnetic waves.

In the past, it might have taken minutes, hours, or even days for news to travel from one part of the United States to another. Today, you can watch and listen to events around the world as they happen! **Broadcasting** *is the use of electromagnetic waves to send information in all directions.* Broadcasting became possible when scientists learned to use radio waves.

FOLDABLES

Make a vertical seven-tab book. Label it as shown. Use it to organize your notes on uses of the different types of electromagnetic waves.

Lesson 3
EXPLORE
615

Radio and Television

How do your radio and television receive information to produce sounds or images? A radio or TV station uses radio waves to carry information. Each radio and TV station sends out radio waves at a certain frequency. The station converts sounds or images into an electric signal. Then, the station produces a **carrier wave**—*an electromagnetic wave that a radio or television station uses to carry its sound or image signals.* The station **modulates,** or varies, the carrier wave to match the electric signal. The signal then gets converted back into images or sounds when it reaches your radio or television.

WORD ORIGIN
modulate
from Latin *modulari*, means "to play or sing"

Figure 8 Amplitude modulation and frequency modulation are two ways of changing a carrier wave to send information.

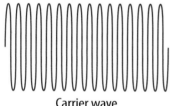
Carrier wave

+

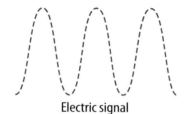
Electric signal

=

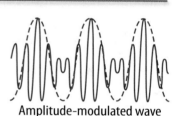
Amplitude-modulated wave

or

Frequency-modulated wave

Two ways a station might change its carrier wave are shown in **Figure 8**. **Amplitude modulation** *(AM) is a change in the amplitude of a carrier wave.* Recall that amplitude is the maximum distance a wave varies from its rest position. **Frequency modulation** *(FM) is a change in the frequency of a carrier wave.* Changes in frequency match changes in sounds or images.

 Key Concept Check How are radio waves used for communication?

Digital Signals

Television stations and some radio stations broadcast digital signals. To understand types of signals, look at the amplitude-modulated wave in **Figure 8.** Notice how the wave changes smoothly from high to low. This type of change is called analog. A digital signal, however, changes in steps. Stations can produce a digital signal by changing different properties of a carrier wave. For example, the station might send the signal as pulses, or a pattern of starting and stopping. The wave might have a code of high and low amplitudes. Sounds and images sent by digital signals are usually clearer than analog signals.

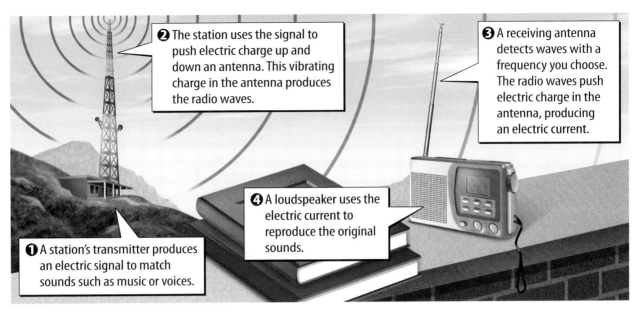

❶ A station's transmitter produces an electric signal to match sounds such as music or voices.

❷ The station uses the signal to push electric charge up and down an antenna. This vibrating charge in the antenna produces the radio waves.

❸ A receiving antenna detects waves with a frequency you choose. The radio waves push electric charge in the antenna, producing an electric current.

❹ A loudspeaker uses the electric current to reproduce the original sounds.

Transmission and Reception

The way a radio station broadcasts a signal is shown in **Figure 9**. A transmitter produces and sends out radio waves. An antenna receives the waves and changes them back to sound. Television transmission is similar, but for cable television, the waves travel through cables designed to carry radio waves.

The wavelength of an AM carrier wave is typically a lot longer than the wavelength of an FM carrier wave. This means AM signals can pass around obstacles and travel farther than FM waves. AM waves also are long enough to reach Earth's upper atmosphere without scattering. FM waves typically have higher frequencies and more energy than AM waves. Therefore, FM waves usually produce better sound quality than AM waves.

Microwaves

You've probably used a microwave oven to heat food. Did you know you also use microwaves when you watch a television show that uses satellite transmission? Like radio waves, microwaves are useful for sending and receiving signals. However, microwaves can carry more information than radio waves because their wavelength is shorter. Microwaves also easily pass through smoke, light rain, snow, and clouds.

Cell Phones

A cell phone company sets up small regions of service called cells. Each cell has a base station with antennae, as shown in **Figure 10**. When you make a call, your phone sends and receives signals to and from the cell tower. Electric circuits in the base station direct your signal to other towers. A signal passes from cell to cell until it reaches the phone of the person you called.

▲ **Figure 9** Radio waves travel from a transmitting station to a radio antenna.

Visual Check What types of energy changes take place during this process?

Figure 10 Cell towers hold antennae from one or more service providers. Each provider positions its antennae toward a different direction to send and receive signals. ▼

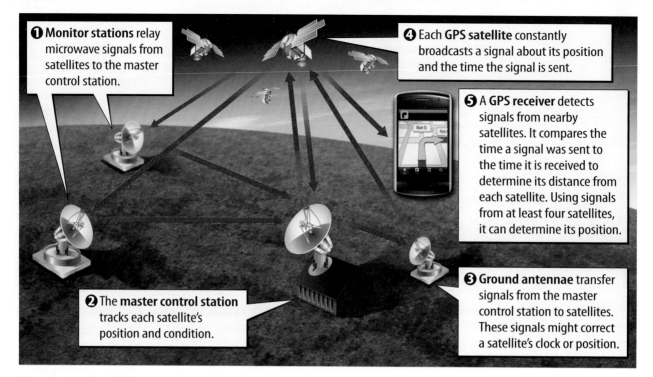

① **Monitor stations** relay microwave signals from satellites to the master control station.

② The **master control station** tracks each satellite's position and condition.

③ **Ground antennae** transfer signals from the master control station to satellites. These signals might correct a satellite's clock or position.

④ Each **GPS satellite** constantly broadcasts a signal about its position and the time the signal is sent.

⑤ A **GPS receiver** detects signals from nearby satellites. It compares the time a signal was sent to the time it is received to determine its distance from each satellite. Using signals from at least four satellites, it can determine its position.

Figure 11 The GPS system includes more than 24 satellites evenly spaced in orbit around Earth. To calculate its position, a receiver needs signals from at least four satellites.

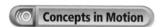
Animation

SCIENCE USE V. COMMON USE
microwave
Science Use a low-frequency, low-energy electromagnetic wave

Common Use appliance used to cook food quickly

ACADEMIC VOCABULARY
analyze
(verb) to study how parts work together

Communication Satellites

Have you seen a sports event or heard a news story from around the world on your TV? Information about the sounds and the images probably traveled by microwaves from a satellite to your TV. Some homes have a satellite dish that receives signals directly from satellites. Satellites send and receive signals similar to the way antennae on Earth do. A transmitter sends microwave signals to the satellite. The satellite can pass the signal to other satellites or send it to another place on Earth.

You can use satellite signals to find directions. *The **Global Positioning System** (GPS) is a worldwide navigation system that uses satellite signals to determine a receiver's location.* As shown in **Figure 11**, at least 24 GPS satellites continually orbit Earth, sending out signals about their orbits and the time. Receivers analyze this information to calculate their location.

 Key Concept Check How are microwaves used for communication?

Infrared Waves

You read in Lesson 2 that vibrating molecules in any matter emit infrared waves. The wavelength of the infrared waves depends on the object's temperature. Hotter objects emit infrared waves with a greater frequency and shorter wavelength than do cooler objects. Scientists have developed technology that produces images showing how much thermal energy a person or thing emits.

Infrared Imaging

Thermal cameras take pictures by detecting infrared waves rather than light waves. They convert invisible infrared waves, or different temperatures, to different colors so that your eyes can interpret the information. Medical professionals use thermal imaging to locate areas of poor circulation in the body. Thermal imaging of a building can identify areas where excessive thermal energy loss occurs.

Can you see in the dark? Night-vision goggles, such as those in **Figure 12,** use infrared waves similar to the way a normal camera uses light. Using the small amount of infrared light present, the goggles produce enough light that objects can be seen easily in the dark.

Imaging Earth

Scientists launch satellites that detect and photograph infrared waves coming from Earth. These infrared images might show variations in vegetation or snowfall. Some images can even show a fire smoldering in a forest before it bursts into flame. The thermal image in **Figure 13** shows the lava flow from a volcano.

 Reading Check What are some ways in which infrared images help humans?

▲ **Figure 12** Infrared waves detected by night-vision goggles make buildings and other objects in a city visible at night.

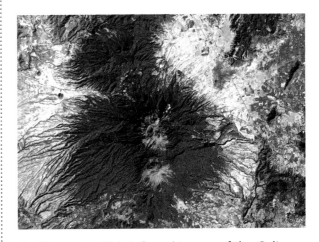

▲ **Figure 13** This infrared image of the Colima Volcano in Mexico shows recent lava flow as red. Notice the volcano's two snow-capped peaks.

Light

The Sun provides most of Earth's light, but light emitted on Earth also has important uses. Think about how difficult driving in a city would be without automobile headlights, street lights, and traffic signals, such as the one in **Figure 14.** Without electric lights, homes and businesses would be dark at night. Televisions, computers, and movie theaters also rely on light. Some forms of communication rely on light that travels through optical fibers—thin strands of glass or plastic that transmit a beam of light over long distances.

Figure 14 A traffic signal is an example of how light is useful. ▶

Inquiry MiniLab 10 minutes

How does infrared imaging work?

Like light waves, infrared waves can be either emitted or reflected by an object. Infrared photography can be used to detect waves that reflect off vegetation.

1. Examine the images. Both show the same location. Compare the true-color visual light image at the top with the false-color infrared image. In the false-color image, red indicates the presence of infrared light.

2. Interpret the images to answer the following questions.

Analyze and Conclude

1. **Analyze Data** What parts of the false-color image show the most infrared light?

2. **Observe** What does the false-color image show about the land in the background?

3. **Key Concept** How could infrared photography be used to identify unhealthy patches of trees in a forest?

Figure 15 A dentist can use an ultraviolet wand to quickly harden adhesives.

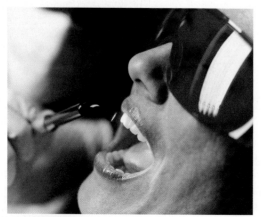

Ultraviolet Waves

Too much exposure to ultraviolet waves can damage your skin, but these same waves also can be useful. Certain materials glow when ultraviolet waves strike them. The materials absorb the energy of the waves and reemit it as light. Credit cards, for example, have invisible symbols stamped on them using this material. If a store clerk holds the card under an ultraviolet lamp, the symbol appears. Scientists also can identify some minerals in rocks by the way they glow under an ultraviolet lamp.

Ultraviolet waves also are useful for killing germs. Just as your skin can be damaged by absorbing the energy of ultraviolet waves, germs also can be damaged. Food manufacturers might use ultraviolet lamps to kill germs in some foods. Campers might use ultraviolet lamps to purify lake water.

Key Concept Check What are some everyday applications of electromagnetic waves?

Medical Uses

In hospitals and clinics, having a germ-free environment might be the difference between life and death. Medical facilities sometimes clean tools and surfaces by bringing them near an ultraviolet lamp. Some air and water purifiers use ultraviolet light to kill germs and reduce the spread of disease.

Ultraviolet light has other medical uses. For example, exposure to ultraviolet waves can help control or cure certain skin problems, such as psoriasis. Several times a week, the patient's skin is exposed to ultraviolet light. The waves carry enough energy to slow the growth of the diseased skin cells.

Another use of ultraviolet waves is shown in **Figure 15.** The dentist is using ultraviolet light to harden an adhesive in just a few seconds. Without ultraviolet light, adhesives might take several minutes to harden.

Fluorescent Lightbulbs

Lightbulbs like the one shown in **Figure 16** use the energy from ultraviolet waves to produce light. Similar to the way a symbol on a credit card glows under ultraviolet light, certain chemicals inside the lightbulb glow when they are exposed to ultraviolet light. When you flip a switch to provide electricity to the bulb, an electric current flows through a gas inside the bulb. The heated gas emits ultraviolet light. This light strikes a chemical that coats the inside of the bulb. The chemical absorbs the energy of the ultraviolet waves and reemits it as light.

 Reading Check How does a fluorescent lightbulb use ultraviolet waves?

▲ **Figure 16** A fluorescent lightbulb uses ultraviolet waves to produce light.

X-Rays

Ultraviolet waves can be dangerous because they can pass through the upper layer of your skin. X-rays have even more energy than ultraviolet waves and can pass through skin and muscle. Although this property makes X-rays dangerous, it also makes them useful. Two important uses of X-rays are security and medical imaging.

Detection and Security

Have you ever had your luggage scanned at an airport, as shown in **Figure 17**? X-rays are useful for security scanning because they can pass through many materials, but metal objects block them. Computers can form images of the contents of luggage by measuring how different materials transmit the X-rays.

Figure 17 Security workers use X-ray scanners to form images of the contents of passenger luggage. ▼

Airport Security

Medical Detection

You might have had an X-ray taken by a dentist or a doctor, similar to that shown in **Figure 18.** X-rays pass through soft parts of your body, but bone stops them. When X-rays strikes photographic film, the film turns dark. Light parts of the film show where bone absorbed the X-rays.

A doctor can obtain even more-detailed images using a computed tomography (CT) scanner. The scanner is an X-ray machine that rotates around a patient, producing a three-dimensional view of the body.

 Reading Check How are a CT scan and a normal X-ray different?

Gamma Rays

Because of their extremely high energy, gamma rays can be used to destroy diseased tissue in a patient. Gamma rays also can be used to diagnose medical conditions. Recall that gamma rays are produced when an atom's nucleus breaks apart or changes. In a procedure called a positron emission tomography (PET) scan, a detector monitors the breakdown of a chemical injected into a patient. The chemical is chosen because it is attracted to diseased parts of the body. The detector can find the location of the disease by detecting the gamma rays emitted by the chemical.

Figure 18 Both X-rays and gamma rays are used in medicine to diagnose and treat diseases.

 Key Concept Check What are some medical uses of electromagnetic waves?

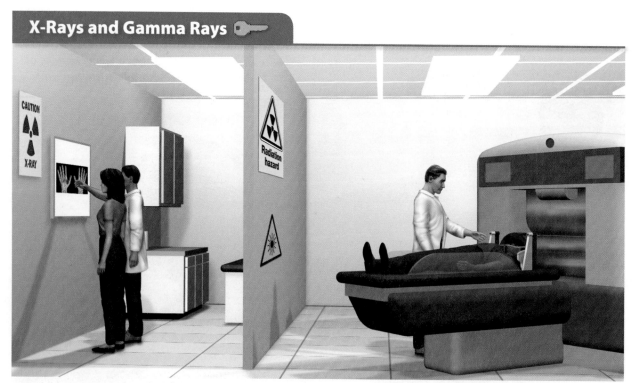

X-Rays and Gamma Rays

Lesson 3 Review

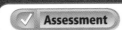

Visual Summary

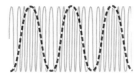

Radio and television stations can broadcast their programming with modulated radio waves.

Satellites send and receive microwave signals to communicate information over long distances.

Because warmer objects emit more infrared waves, scientists can study heat-related changes on Earth's surface, such as weather and volcanic activity.

FOLDABLES

Use your lesson Foldable to review the lesson. Save your Foldable for the project at the end of the chapter.

What do you think NOW?

You first read the statements below at the beginning of the chapter.

5. Thermal images show differences in the amount of energy people or objects give off.

6. When you call a friend on a cell phone, a signal travels directly from your phone to your friend's phone.

Did you change your mind about whether you agree or disagree with the statements? Rewrite any false statements to make them true.

Use Vocabulary

1. The wave that is changed to carry a radio or television signal is a(n) _____.

2. **Use the term** *broadcasting* in a complete sentence.

Understand Key Concepts

3. Which of the following changes in an AM radio wave?
   A. amplitude   C. speed
   B. frequency   D. wavelength

4. **Explain** why X-rays are less harmful than infrared waves in medical imaging.

5. **Describe** how microwaves are useful for communication.

6. **Identify** an everyday application of ultraviolet waves.

Interpret Graphics

7. **Interpret** Which type of modulation does the radio wave below have? Do you think a station could broadcast this signal for hundreds of kilometers? Explain.

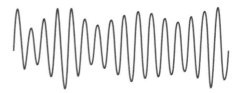

8. **Identify** Copy and fill in a graphic organizer like the one below to identify several uses of electromagnetic waves.

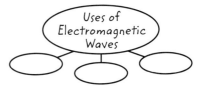

Critical Thinking

9. **Predict** the effect on human life if the ozone layer, which blocks harmful ultraviolet waves, becomes thinner or develops more holes.

3 class periods

Materials

miscellaneous supplies for making an exhibit—poster board, markers, balsa wood, masking tape, etc.

Safety

Design an Exhibit for a Science Museum

Scientists learn about distant regions of the universe by detecting electromagnetic waves from deep space. Doctors use electromagnetic waves to find and treat diseases. Anytime you use a cell phone or watch a television program, you use electromagnetic waves. Even though people use electromagnetic waves every day, many people don't know much about the electromagnetic spectrum. Your assignment is to help change that.

Question

What sort of science museum exhibit can you design to educate others about electromagnetic waves?

Procedure

1. Read and complete a lab safety form.

2. Brainstorm with your partner(s) about different types of exhibits you enjoy when visiting museums (sports, science, history, music) or zoos. What makes them fun? Make a list of your ideas titled *Exhibits* in your Science Journal.

3. Discuss with your partner(s) ideas, facts, or concepts you remember from your visits to museums or zoos. What was it about your experience that made this information memorable? Make a list of these qualities titled *Memorable*.

4. Determine with your class which part of the electromagnetic spectrum your exhibit will describe.

5. Research your assigned part of the electromagnetic spectrum using your textbook, reliable sources on the Internet, and other reference materials. You also might interview people. In addition to specifics about your topic, your exhibit should provide a general overview of the electromagnetic spectrum.

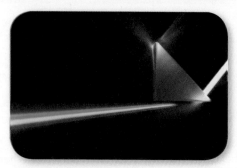

Chapter 17
EXTEND

6. Plan your exhibit. Where will the information portion be? Is the exhibit interactive? Will people do something when they visit your exhibit? Will any materials be used while people visit? Are your ideas safe?

7. Gather the supplies you need, and then construct your exhibit.

8. Invite your classmates to evaluate your exhibit. Have them write on an index card what they enjoyed and what they remembered after leaving your exhibit. Ask them to write questions or suggestions they might have about your topic or exhibit.

Analyze and Conclude

9. **Analyze Data** Read the comments made on the index cards by your classmates. Organize this information into three lists: things people liked, things people learned, and questions or suggestions people made.

10. **Draw Conclusions** Use the feedback lists to evaluate your display. Were you able to create an informative and fun exhibit? Is there anything you would change or improve based on the feedback?

11. **The Big Idea** How did you describe and use electromagnetic waves in your exhibit?

Lab Tips

☑ Consider your lists of exhibits and memorable qualities carefully.

☑ When preparing your exhibit, think about your audience. Make your exhibit understandable and entertaining for someone your age.

☑ Quality counts. Take time to make a neat and error-free exhibit.

Communicate Your Results

Work with your classmates to place your exhibits in a public area of the school. Invite other students to visit your class museum.

 **Extension**

Research a topic you found interesting when you visited your classmates' exhibits. Design an experiment, write a paper, or make a poster to explore or explain the topic.

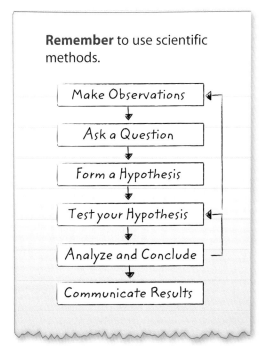

Lesson 3 EXTEND

Chapter 17 Study Guide

 Electromagnetic waves move through matter and space. They have different frequencies, wavelengths, and energy. Uses include communication and medical imaging.

Key Concepts Summary

Vocabulary

Lesson 1: Electromagnetic Radiation

- An accelerating electric charge produces an **electromagnetic wave.**
- Electromagnetic waves travel through space at 300,000 km/s. Properties include their wavelengths, their frequencies, and the amount of **radiant energy** each wave carries.

electromagnetic wave p. 601

radiant energy p. 601

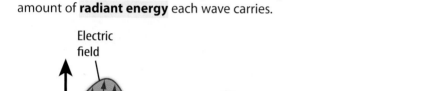
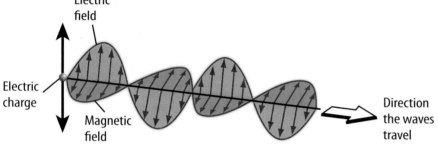

Lesson 2: The Electromagnetic Spectrum

- The **electromagnetic spectrum** is the entire range of wavelengths and frequencies of electromagnetic waves.
- Electromagnetic waves differ in their wavelengths, frequencies, and energy. **Radio waves** have the longest wavelengths, the lowest frequencies, and the lowest energy. **Gamma rays** have the shortest wavelengths, the highest frequencies, and the highest energy.

electromagnetic spectrum p. 609

radio wave p. 610

microwave p. 610

infrared wave p. 610

ultraviolet wave p. 611

X-ray p. 611

gamma ray p. 611

Lesson 3: Using Electromagnetic Waves

- Electromagnetic waves are used in radio and television **broadcasting,** and in cell phone communication.
- Using light to see and microwaves to cook food are everyday applications of electromagnetic waves. Listening to radio and watching television require devices that use electromagnetic waves.
- Doctors use X-rays to identify broken bones or damaged teeth. Gamma rays can be used to diagnose or treat diseases.

broadcasting p. 615

carrier wave p. 616

amplitude modulation p. 616

frequency modulation p. 616

Global Positioning System p. 618

Study Guide

Review
- Personal Tutor
- Vocabulary eGames
- Vocabulary eFlashcards

Chapter Project

Assemble your lesson Foldables as shown to make a Chapter Project. Use the project to review what you have learned in this chapter.

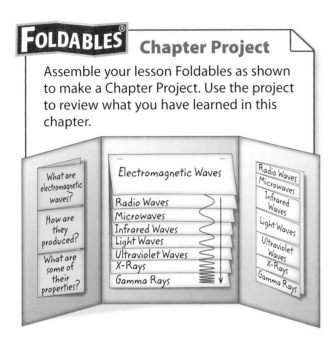

Use Vocabulary

1. The energy carried by electromagnetic waves is _____.
2. A transverse wave that is able to move through space is a(n) _____.
3. The type of electromagnetic wave that causes sunburn is a(n) _____.
4. The electromagnetic wave that has the highest energy and is therefore most dangerous to humans is a(n) _____.
5. A television signal travels from one place to another by a(n) _____ that is modulated.
6. An FM station changes its carrier wave by _____.

Link Vocabulary and Key Concepts

Interactive Concept Map

Copy this concept map, and then use vocabulary terms from the previous page to complete the concept map.

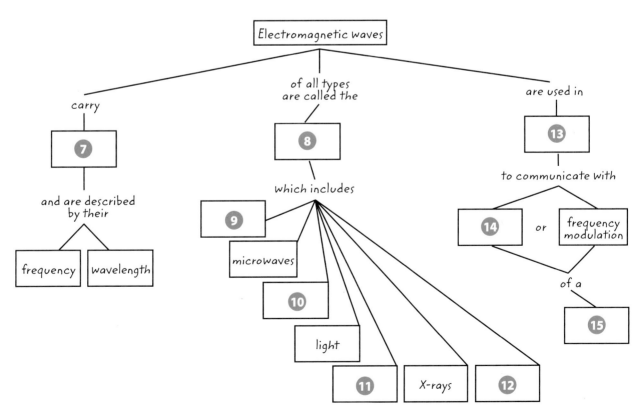

Chapter 17 Review

Understand Key Concepts

1. What happens to an electromagnetic wave as it passes from space to matter?
   A. It changes frequency.
   B. It changes wavelength.
   C. It slows down.
   D. It speeds up.

2. Which property of the carrier wave shown below has been modulated?

   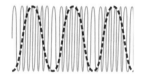

   A. amplitude
   B. energy
   C. frequency
   D. speed

3. Which type of electromagnetic waves has the longest wavelengths?
   A. infrared waves
   B. microwaves
   C. radio waves
   D. ultraviolet waves

4. Which type of electromagnetic waves does your body emit?
   A. infrared waves
   B. microwaves
   C. radio waves
   D. ultraviolet waves

5. Which three types of electromagnetic waves carry most of the Sun's energy that strikes Earth?
   A. infrared waves, light, ultraviolet waves
   B. ultraviolet waves, light, microwaves
   C. light, ultraviolet waves, radio waves
   D. X-rays, infrared waves, gamma waves

6. What happens to a wave if its frequency decreases?
   A. Its amplitude increases.
   B. Its energy decreases.
   C. Its speed increases.
   D. Its wavelength decreases.

7. Which list of electromagnetic waves is in the correct order from highest to lowest energy?
   A. gamma rays, radio waves, infrared waves, microwaves
   B. ultraviolet waves, gamma rays, X-rays, light
   C. light, infrared waves, microwaves, radio waves
   D. X-rays, gamma rays, ultraviolet waves, light

8. The image below most likely was produced using which type of electromagnetic wave?

   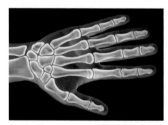

   A. gamma rays
   B. light
   C. microwaves
   D. X-rays

9. Which type of electromagnetic wave causes a chemical to glow in a fluorescent lightbulb?
   A. gamma waves
   B. infrared waves
   C. radio waves
   D. ultraviolet waves

Chapter Review

Assessment
Online Test Practice

Critical Thinking

10 Classify Look around your home, school, and community. Make a list of different devices that use electromagnetic waves. Beside each device, write the type of electromagnetic wave the device uses.

11 Sequence the types of electromagnetic waves according to their ability to penetrate matter.

12 Synthesize The waves in the diagram below represent the colors yellow, blue, and red. Describe which color is represented by waves A, B, and C, and explain how you made your choice.

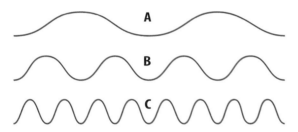

13 Apply You are in a dark room. Are there any electromagnetic waves present? Why or why not?

14 Explain how you can determine the frequency of an electromagnetic wave moving through space if you know the wavelength.

15 Assess New Zealand has one of the highest rates of skin cancer in the world. In terms of electromagnetic waves, what might be causing the disease?

Writing in Science

16 Write A classmate can't understand what electromagnetic waves have to do with his or her everyday life. Write a note at least five sentences long to your friend explaining why understanding electromagnetic waves is important and how the information might apply to everyday life.

REVIEW THE BIG IDEA

17 What are electromagnetic waves? How do they travel outward from their source?

18 The photos below show that the Sun emits different types of electromagnetic waves. How can you describe the main types of electromagnetic waves the Sun emits? What are some ways in which people use electromagnetic waves?

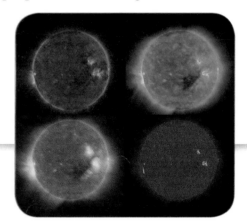

Math Skills

Math Practice

Solve One-Step Equations

19 What is the wavelength of an electromagnetic wave with a frequency of 145,000 Hz?

20 What is the wavelength of an electromagnetic wave with a frequency of 165,000 Hz?

21 An electromagnetic wave has a wavelength of 0.65 km. What is its frequency?

22 An electromagnetic wave has a wavelength of 0.45 km. What is its frequency?

23 What is the wavelength of an electromagnetic wave with a frequency of 225,000 Hz?

Standardized Test Practice

Record your answers on the answer sheet provided by your teacher or on a sheet of paper.

Multiple Choice

Use the figures to answer questions 1 and 2.

Electromagnetic wave

Rope wave

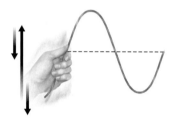

1. The figures show an electromagnetic wave and a wave in a rope. Which characteristic do these two waves have in common?
   A They are both transverse waves.
   B They are both longitudinal waves.
   C They both have electric and magnetic fields.
   D They both have compressions and rarefactions.

2. How do these waves differ?
   A One has crests, while the other has troughs.
   B One can travel through empty space, while the other cannot.
   C One transmits energy, while the other transmits matter.
   D One has an electric field, while the other has a magnetic field.

3. Which forms an electromagnetic wave when it vibrates?
   A an electron
   B a carrier wave
   C a photon
   D an X-ray

4. Which shows the correct order from lowest to highest energy for different electromagnetic waves?
   A light → X-ray → microwave
   B gamma ray → light → radio wave
   C X-ray → microwave → ultraviolet wave
   D radio wave → microwave → gamma ray

Use the figure to answer questions 5 and 6.

Violet   Blue   Green   Yellow   Orange   Red

5. The figure lists different colors of light. Based on the sequence shown, what could the arrow represent?
   A increasing energy
   B increasing frequency
   C increasing speed
   D increasing wavelength

6. What do all these colors of light have in common?
   A their energy
   B their frequency
   C their speed
   D their wavelength

630 • Chapter 17 Standardized Test Practice

Standardized Test Practice

Use the figure to answer questions 7 and 8.

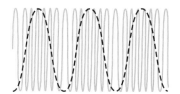

7 The figure shows a carrier wave used to broadcast radio signals. What is modulated in this wave in order to carry information?

   A amplitude
   B crest height
   C frequency
   D loudness

8 Which is true of this type of signal wave?

   A It can easily pass around obstacles.
   B It can be seen with the unaided eye.
   C It can travel far by bouncing off particles in the atmosphere.
   D It can have wavelengths that are less than a meter long.

9 Which is the number of satellites needed to calculate the position of a Global Positioning System (GPS) receiver on Earth?

   A 2
   B 4
   C 6
   D 8

Constructed Response

10 How does the speed of an electromagnetic wave relate to its frequency and its wavelength? If the speed of light is 300,000 km/s through empty space, how would you calculate the wavelength of light that has a frequency of 150,000 Hz?

Use the figure to answer questions 11 and 12.

11 Describe how electromagnetic waves A, B, and C differ (if at all) in amplitude, wavelength, frequency, and energy?

12 Choose three types of electromagnetic radiation which would fit the characteristics of the waves shown. Identify which wave could represent each type and why.

13 What kind of information does an infrared image taken from a satellite give about Earth? How might this kind of information be used?

NEED EXTRA HELP?													
If You Missed Question...	1	2	3	4	5	6	7	8	9	10	11	12	13	14
Go to Lesson...	1	1	1	2	2	2	3	3	3	3	1	2	2	3

Chapter 18

Light

THE BIG IDEA How does matter affect the way you perceive and use light?

Inquiry Are they real?

The tiny flowers seem real, but they are not. This is a close-up photo of a chrysanthemum flower covered with water drops. The tiny flower images are reflections in the water drops.

- What causes images of the flower to appear in the water drops?
- Why do reflections appear in the water but not on the flower?
- How does matter affect the way you perceive light?

Get Ready to Read

What do you think?
Before you read, decide if you agree or disagree with each of these statements. As you read this chapter, see if you change your mind about any of the statements.

1. Both the Sun and the Moon produce their own light.
2. The color of an object depends on the light that strikes it.
3. Mirrors are the only surfaces that reflect light.
4. The image in a mirror is always right-side up.
5. A prism adds color to light.
6. When light moves from air into water or glass, it travels more slowly.
7. A laser will burn a hole through your skin.
8. Telephone conversations can travel long distances as light waves.

ConnectED Your one-stop online resource

connectED.mcgraw-hill.com

- Video
- Audio
- Review
- Inquiry
- WebQuest
- Assessment
- Concepts in Motion
- Multilingual eGlossary

Lesson 1

Light, Matter, and Color

Reading Guide

Key Concepts 🗝
ESSENTIAL QUESTIONS

- What are some sources of light, and how does light travel?
- What can happen to light that strikes matter?
- Why do objects appear to have different colors?

Vocabulary

light p. 635
reflection p. 635
transparent p. 636
translucent p. 636
opaque p. 636
transmission p. 637
absorption p. 637

 Multilingual eGlossary

 Video BrainPOP®

Inquiry Why So Colorful?

On a clear day, it's exciting to watch colorful hot-air balloons launch into the blue sky. What makes the balloons so colorful? Did you ever wonder how you are able to see all the different colors?

634 Chapter 18
ENGAGE

Inquiry Launch Lab

10 minutes

How can you make a rainbow?

After it rains, you might see a rainbow stretching across the sky. How can sunlight change into so many different colors? How can you make a rainbow of colors by shining light through a prism?

1. Read and complete a lab safety form.
2. Use **tape** to attach a sheet of **white paper** to a wall.
3. In a darkened room, stand about 1 m from the paper. Hold a **prism** in front of a **flashlight,** as shown in the photo. Turn on the flashlight. Shine the light through the prism onto the paper.
4. Adjust the prism until you see bands of color on the paper. Turn the prism, and have your partner trace the bands of color on the paper with **colored pencils.** Record your observations in your Science Journal. Turn the prism two more times, and trace the colors.
5. Switch places with your partner, and repeat steps 3–4.

Think About This

1. What is the sequence of colors in the bands of light in your drawings? Is it the same in all the drawings?
2. **Key Concept** How do you think the prism affected the white light from the flashlight?

What is light?

Suppose you are taking a tour of a cave below Earth's surface and the lights go out. Would you see anything? No! You could not see anything because there would be no light for your eyes to perceive. **Light** is *electromagnetic radiation that you can see*. Electromagnetic radiation has wave properties and particle properties. A particle of electromagnetic radiation is called a photon. The frequency of a light wave depends on the amount of energy carried by a photon of light. Light waves can carry this energy through space and some matter.

Sources of Light

You can see an object if it is luminous or if it is illuminated. Luminous objects, such as the campfire in **Figure 1,** release, or emit, light. The Sun is luminous because it is made of hot, glowing gases. A traffic light and a firefly also are luminous objects. Luminous objects are sources of light. We also can see illuminated objects, such as the camper and trees in **Figure 1.** However, they do not emit light and are not sources of light. The Moon might appear bright, but it is just a rocky sphere. You see the Moon because it reflects light from the Sun. **Reflection** *is the bouncing of a wave off a surface*. In a dark cave, there is no light to reflect off objects so you cannot see anything.

Key Concept Check What are some sources of light?

REVIEW VOCABULARY

electromagnetic radiation
energy carried through space or matter by electromagnetic waves

Figure 1 You see the campfire because it is luminous. You see other objects in the picture because light from the fire illuminates them.

How Light Travels

Light travels as waves moving away from a source. Scientists often describe these waves as countless numbers of light rays spreading out in all directions from a source. You can model light rays by shining a light through a comb, as shown in **Figure 2.**

What are the spaces between the rays in **Figure 2?** They are shadows or places with less light. The shadows show that light normally travels in a straight line. However, objects in its path can cause light to change direction. It also can spread out slightly as it moves through a small opening. Sometimes, the effect is difficult to see because light waves are so small.

 Key Concept Check How does light travel?

Light and Matter

What can you see through your classroom window? Would you still be able to see anything if the blinds were closed? How do different types of matter affect light?

You can see objects clearly through air, clean water, plain glass, and some plastics. *A material that allows almost all the light that strikes it to pass through and form a clear image is* **transparent.** The unfrosted parts of the window in **Figure 3** are transparent.

Light also passes through the frosted parts of the window, but clear images do not form. *A material that allows most of the light that strikes it to pass through and form a blurry image is* **translucent.** Plastics with textured surfaces also are examples of translucent materials.

A material through which light does not pass is **opaque.** No light passes through the chairs or the table in **Figure 3**. Wood and metal are examples of opaque materials.

 Key Concept Check What can happen to light that strikes matter?

▲ **Figure 2** Light spreads out as it moves away from a source. An object in the path of light forms shadows where there is less light.

Figure 3 Different amounts of light can pass through the transparent and translucent parts of this window. The opaque walls and table do not allow light to pass through them. ▼

Figure 4 The window transmits, absorbs, and reflects light.

Visual Check Which materials in the photograph are transparent? Which are opaque?

Transmission of Light

You just read that light passes through transparent and translucent objects. *The passage of light through an object is called* **transmission.** The girls and you can see objects through the window in **Figure 4** because the window transmits light. A luminous object or an illuminated object on one side of the window is visible on the other side of the window. Also, the energy carried by the light waves from these objects can pass through the window.

Absorption of Light

Imagine standing near a window on a spring day. The transparent window transmits some sunlight, and it lands on you. If you touch the window, it might feel warm. Some of the energy in sunlight stays inside the window. *The transfer of energy by a wave to the medium through which it travels is called* **absorption.** The energy causes atoms in the material to vibrate faster, increasing the temperature of the material. All materials absorb some of the light that strikes them. The window feels warm because it absorbs some of the sunlight's energy.

Reflection of Light

When you look at a pane of glass, you sometimes can see an *image* of yourself. Light bounces off you, strikes the glass, and bounces back to your eye. Recall that the bouncing of a wave off a surface is called reflection.

Look again at the window in **Figure 4.** Notice that you can see the transmission and reflection of light. You cannot see it, but some of the light also is absorbed. Most types of matter interact with light in a combination of ways.

WORD ORIGIN

transmission
from Latin *transmissionem,* means "sending over or across"

FOLDABLES
Make a vertical trifold book and label it as shown. Use it to organize your notes on the sources of light, light and matter, and the relationship between light and color.

Light Sources

Light & Matter

Light & Color

ACADEMIC VOCABULARY

image
(noun) a figure, picture, or representation of an object

Inquiry MiniLab
15 minutes

What color is that?
Have you ever noticed that objects seem to change color when you put on sunglasses? What causes this to happen?

Object	Filter			
	None	Red	Blue	Green

1. Read and complete a lab safety form.
2. Copy the data table into your Science Journal. You can add lines as needed.
3. Write *PHYSICS!* on **white paper** as shown here using **red** and **blue markers.**
4. Choose at least **4 different-colored objects.** Observe the word *PHYSICS!* and the objects with no filter and then through **red, blue,** and **green filters.** Record your observations in your Science Journal.

Analyze and Conclude

1. **Contrast** your observations with and without filters.
2. **Key Concept** Explain why the objects you observed appeared to change color when you used a filter.

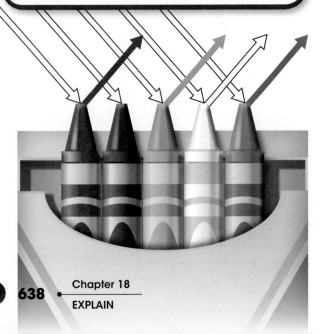

Light and Color

Recall that visible light is electromagnetic radiation with wavelengths of all colors of the rainbow. The longest wavelengths of light appear red. Violet has the shortest wavelengths. Other colors of light have different wavelengths. White light is a mixture of all wavelengths of light. How does this account for the colors you might see?

Colors you see result from wavelengths of light that enter your eyes. When you look at a luminous object such as a campfire, you see the colors emitted by the fire and glowing logs. What happens when you look at an illuminated object? That depends on whether the object is opaque, transparent, or translucent.

Opaque Objects

Suppose white light strikes a box of crayons, as shown in **Figure 5.** Each crayon absorbs all wavelengths of light except its color. For example, the green crayon absorbs all colors except green. The green wavelengths reflect back into your eyes, and you see green. The red crayon absorbs all colors except red, and it reflects red. The black crayon absorbs all colors. The color of an opaque object is the color of light it reflects.

What do you think would happen to the colors reflected by the crayons if red light, instead of white light, were shining on them? Would the green crayon still appear green? No. It would absorb the red light, but there would be no green light to reflect. The green crayon would appear black. The blue crayon also would appear black. The red crayon and the white crayon would appear red because they would reflect red light. The color you see always depends on the color of light that the object reflects.

Figure 5 The color of an object is the color of light that reflects off the object.

Concepts in Motion Animation

638 • Chapter 18
EXPLAIN

Transparent and Translucent Objects

Absorption, transmission, and reflection also explain the color of transparent and translucent objects. For example, suppose white light, such as sunlight, shines through a piece of blue glass. The glass absorbs all wavelengths of light except blue. The blue wavelengths pass through the glass to your eyes. If the blue glass is translucent, it still only transmits blue light, but the image is blurry. The color of a transparent or translucent object is the color it transmits.

What color would you see if you held a piece of red cellophane, or a red filter, in front of each crayon in **Figure 5**? Remember that the color you see depends on the colors in the light source, the absorbed colors, and the colors that reach your eyes.

Combining Colors

Have you ever mixed several colors of watercolor paints to get just the shade you want? You know that you can make many different shades from a few basic colors. If you mix too many colors, you get black! Why does that happen?

Combining Pigments Each color of paint in a set of watercolors contains different pigments, or dyes. Each pigment absorbs some colors of light and reflects other colors. Mixing pigments produces many different shades as certain wavelengths are absorbed and fewer colors are reflected to your eyes. As you add each color of pigment, the mixture gets darker and darker because more colors are absorbed. Cyan, magenta, and yellow are the primary pigments. Combining equal amounts of these pigments makes black, as shown in the center of the artist's palette in **Figure 6**.

Combining Colors of Light Red, green, and blue are the primary light colors. If you shine equal amounts of red light, green light, and blue light at a white screen, each color reflects to your eyes. Where two of the colors overlap, both wavelengths reflect to your eyes and you see a third color. Where the three colors overlap, all colors reflect and you see white light, also shown in **Figure 6**.

 Key Concept Check Why do objects appear to have different colors?

Figure 6 Colors of pigment combine by subtracting wavelengths. Colors of light combine by adding wavelengths.

Visual Check What is the result of adding any two primary pigments?

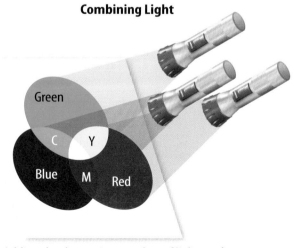

Combining Pigments

Each primary pigment subtracts color by absorption. Mixing all three pigments equally produces black.

Combining Light

Adding the three primary colors of light produces the colors of the primary pigments and white.

Lesson 1 Review

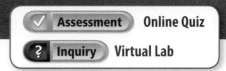

Visual Summary

The amount of light that can pass through a material determines whether it is transparent, translucent, or opaque.

Reflection is the bouncing of waves off a surface.

The color of an opaque object is the color of light that reflects off the object.

FOLDABLES

Use your lesson Foldable to review the lesson. Save your Foldable for the project at the end of the chapter.

What do you think NOW?

You first read the statements below at the beginning of the chapter.

1. Both the Sun and the Moon produce their own light.
2. The color of an object depends on the light that strikes it.

Did you change your mind about whether you agree or disagree with the statements? Rewrite any false statements to make them true.

Use Vocabulary

1. **Compare and contrast** how an opaque object and a transparent object affect light.

2. The passage of light through an object is called _____.

3. An electromagnetic wave that you can see is _____.

Understand Key Concepts

4. Which indicates that light travels in straight lines?
   A. colors   C. shadows
   B. pigments   D. waves

5. **Contrast** how white light interacts with a blue book and a blue stained glass window.

6. **Describe** three things that can happen when light strikes an object.

Interpret Graphics

7. **Explain** why the overlap region of the filters at the right looks black when you look through the filters toward a white light.

8. **Summarize** Copy the graphic organizer below. Fill in three ways you can describe a material according to whether light can pass through it.

Critical Thinking

9. **Construct** a diagram showing what happens when rays of white light strike a yellow, opaque object.

Color!

SCIENCE & SOCIETY

There's more to it than meets the eye.

Do you have a favorite color? If you had to explain why it's your favorite, what would you say? You know that an object's color is due to the wavelengths of light it reflects, but you might be surprised to learn that color is more than something you see. Color can affect your mood, your emotions, and even your creativity at school or work!

Color psychology is the study of the effect of color on the way people think and behave. For example, the color blue is associated with peacefulness. It tends to make people feel calm. Studies have shown that blue rooms promote creativity.

▲ This waiting room is designed to be soothing.

What words would you use to describe a bright yellow room? Would you say the room is sunny and cheerful? People often connect happiness with yellow and nature with green. Green, like blue, tends to calm people. Hospitals often use green in waiting rooms to reduce people's anxiety. What about orange? Many people connect orange with energy and warmth.

Facts about how color affects moods can be fun to learn. However, for some people, it is their business. Industrial psychologists study ways to maximize workers' productivity and safety. Many industrial psychologists pay close attention to the colors used in different areas of the workplace. Those who work in advertising and merchandising choose specific colors for product packages that connect the product inside with certain ideas or emotions.

So, what does your favorite color mean to you? Perhaps it changes depending on your mood or experiences. Regardless of what your favorite color is, it is interesting to know that there is science behind colors chosen for many places and products used daily.

▲ Do you think that blue calms people because it makes them think about the ocean or the sky?

Red makes people think of energy and excitement. That's why it's a popular color for sports cars.

It's Your Turn

RESEARCH AND REPORT Find out more about how a person's culture might influence the way the person perceives colors. Make a poster to share what you learn with your classmates.

Lesson 1 EXTEND

Lesson 2

Reading Guide

Key Concepts
ESSENTIAL QUESTIONS

- How does light reflect from smooth surfaces and rough surfaces?
- What happens to light when it strikes a concave mirror?
- Which types of mirrors can produce a virtual image?

Vocabulary
law of reflection p. 643
regular reflection p. 644
diffuse reflection p. 644
concave mirror p. 645
focal point p. 645
focal length p. 645
convex mirror p. 646

g Multilingual eGlossary

Reflection and Mirrors

Inquiry Why do they appear so strange?

When you look in a mirror in your home, your image is very much like you. But these people are looking at their images in a carnival's fun-house mirror. Some parts of their images are much larger than they should. Other parts are much smaller. What causes the images to appear so strange?

Inquiry Launch Lab

10 minutes

How can you read a sign behind you?

Mirrors aren't just for looking at your reflection. You can use them to see around corners, over and under obstacles, and inside things.

1. Read and complete a lab safety form.
2. Use a **marker** to write *CAT* about 5 cm tall on an **index card.** Use **tape** to attach the card to a wall.
3. Stand with your back toward the wall. Use a **mirror** to look at the letters on the card. Sketch what you see in your Science Journal.
4. Repeat step 3, but this time use **an additional mirror** to look at the letters in the first mirror.

Think About This

1. Draw a simple diagram showing the path of light from the card to your eye for each setup.
2. **Key Concept** How did one reflection in a mirror affect the letters? How did two reflections affect the letters?

Reflection of Light

In Lesson 1, you read that reflection is the bouncing of a wave off a surface. Suppose you toss a tennis ball against a wall. If you throw the ball straight toward the wall, it will bounce back to you. Where on the wall would you throw the ball so that a friend standing to your left could catch it? You would throw it toward a point on the wall halfway between you and your friend.

Law of Reflection

Like the tennis ball, light behaves in predictable ways when it reflects. Scientists often model the path of light using straight arrows called rays. The rays in **Figure 7** show how light reflects. An imaginary line perpendicular to a reflecting surface is called the normal. The light ray moving toward the surface is the incident ray. The light ray moving away is the reflected ray.

Notice the angle formed where an incident ray meets a normal. This is the angle of incidence. A reflected ray forms an identical angle on the other side of the normal. This angle is the angle of reflection. According to the **law of reflection**, *when a wave is reflected from a surface, the angle of reflection is equal to the angle of incidence.*

Reading Check How do the reflected and incident rays differ?

Figure 7 If you know a light ray's angle of incidence, you can predict its angle of reflection.

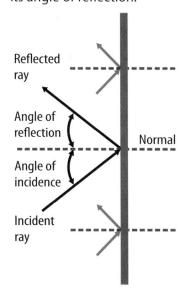

Visual Check If the angle of incidence is 40°, what is the angle of reflection?

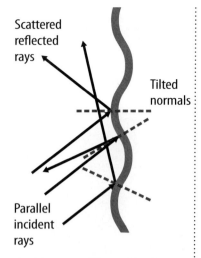

▲ Figure 8 🔑 No clear image forms when light reflects off a rough surface.

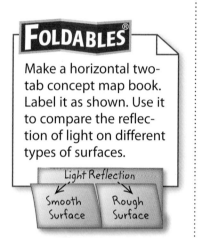

Regular and Diffuse Reflection

You see objects when light reflects off them into your eyes. Why can you see your reflection in smooth, shiny surfaces but not in a piece of paper or a painted wall? The law of reflection applies whether the surface is smooth or rough.

Reflection of light from a smooth, shiny surface is called **regular reflection.** Look again at **Figure 7** on the previous page. The three incident rays and the three reflected rays all are parallel. You see a sharp image when parallel rays reflect into your eyes. When light strikes an uneven surface, as in **Figure 8,** the angle of reflection still equals the angle of incidence at each point. However, rays reflect in different directions. *Reflection of light from a rough surface is called* **diffuse reflection.**

 Key Concept Check How does light reflect from smooth surfaces and rough surfaces?

Mirrors

Any surface that reflects light and forms images is a mirror. The type of image depends on whether the reflecting surface is flat or curved.

Plane Mirrors

The word *plane* means "flat," so a plane mirror has a flat reflecting surface. **Figure 9** shows how an image from a plane mirror forms Only a few rays are shown, but the number of rays involved is infinite. Notice that the image formed is the same size as the object. However, it is a virtual image. A virtual image is an image of an object that your brain perceives to be in a place where the object is not.

If you look in a plane mirror and raise your right hand, your image raises its left hand. However, up and down are not reversed.

Figure 9 🔑 A plane mirror forms a virtual image.

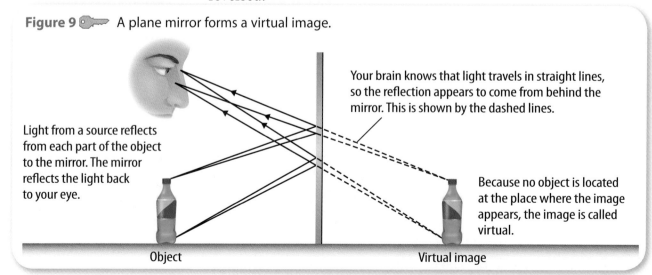

Inquiry MiniLab

15 minutes

Where is the image in a plane mirror?

A plane mirror forms a virtual image. Where is it, compared to the real object?

1. Read and complete a lab safety form.
2. Prop a sheet of **clear acrylic** vertically between **two books**.
3. Place a **candle** 15 cm in front of the acrylic. Use a **safety match** to light the candle in a slightly darkened room.
4. Place an **unlit candle** behind the acrylic. Look through the acrylic. Move the unlit candle to align with the reflection of the lit candle. Observe from several locations to the right and the left of center. Record your observations in your Science Journal.
5. Using a **ruler,** measure from the unlit candle to the acrylic. Record the distance.
6. Repeat steps 4–5 several times with the lit candle different distances from the acrylic.

Analyze and Conclude

1. **Compare** the distances of the unlit candle and the lit candle from the acrylic.
2. **Draw** a diagram of your setup. Use arrows and labels to show reflection of light and how your brain perceives a virtual image.
3. **Key Concept** Describe and explain the reflection you saw.

Concave Mirrors

Not all mirrors are flat. *A mirror that curves inward is called a* **concave mirror**, like the mirror in **Figure 10**. A line perpendicular to the center of the mirror is the optical axis. The law of reflection determines the direction of reflected rays in **Figure 10**. When rays parallel to the optical axis strike a concave mirror, the reflected rays converge, or come together.

Focal Point Look again at **Figure 10**. Notice the point where the rays converge. *The point where light rays parallel to the optical axis converge after being reflected by a mirror or refracted by a lens is the* **focal point.** In the next lesson, you will read about lenses. Imagine that a concave mirror is part of a hollow sphere. The focal point is halfway between the mirror and the center of the sphere. *The distance along the optical axis from the mirror to the focal point is the* **focal length.** The lesser the curve of a mirror, the longer its focal length. The position of an object compared to the focal point determines the type of image formed by a concave mirror.

If you reverse the direction of the arrows in **Figure 10,** you can see how a flashlight works. The reflector behind the bulb is a concave mirror. The bulb is at the focal point. Light rays from the bulb strike the mirror and reflect as parallel rays.

Key Concept Check How does a concave mirror reflect light?

WORD ORIGIN

concave
from Latin *concavus*, means "hollow"

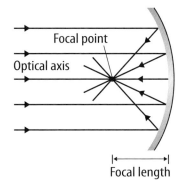

Figure 10 Light rays that strike a concave mirror converge.

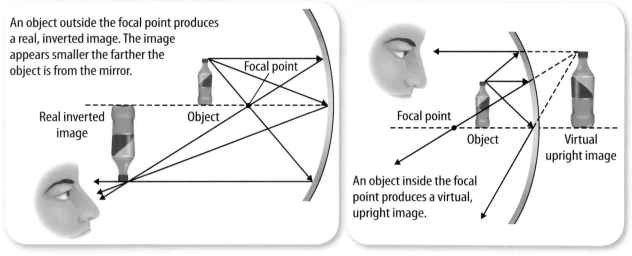

An object outside the focal point produces a real, inverted image. The image appears smaller the farther the object is from the mirror.

An object inside the focal point produces a virtual, upright image.

▲ **Figure 11** A concave mirror can produce a real or a virtual image.

▲ **Figure 12** The image flips as the spoon moves away from the object.

Figure 13 The image in a convex mirror looks smaller than the object. ▼

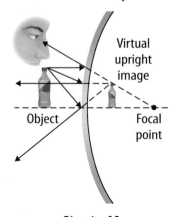

Types of Images When you look at ray diagrams, think about what you would see if the rays reflected into your eyes. Remember that your brain only perceives where the rays appear to come from, not their actual paths. As shown in **Figure 11,** the image a concave mirror forms depends on the object's location relative to the focal point. The image is virtual if the object is between the focal point and the mirror. The image is real if the object is beyond the focal point. A real image is one that forms where rays converge. No image forms if the object is at the focal point.

Suppose you look at your reflection in the bowl of a shiny spoon. If your face is outside the spoon's focal point, your image is inverted, or upside down, as shown in **Figure 12.** Your image disappears when your face is at the focal point. When your face is inside the focal point, the image is upright.

Convex Mirrors

Have you ever seen a large, round mirror high in the corner of a store? The mirror enables store personnel to see places they cannot see with a plane mirror. *A mirror that curves outward, like the back of a spoon, is called a* **convex mirror.** Light rays diverge, or spread apart, after they strike the surface of a convex mirror. The dashed lines in **Figure 13** model how your brain interprets these rays as coming from a smaller object behind the mirror. Therefore, a convex mirror always produces a virtual image that is upright and smaller than the object.

As you just read, convex mirrors and plane mirrors form only virtual images. However, concave mirrors can form both virtual images and real images.

Key Concept Check Which types of mirrors can form virtual images?

Lesson 2 Review

Assessment · Online Quiz

Visual Summary

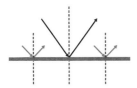

The angle of reflection equals the angle of incidence, even if a surface is rough.

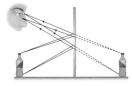

The reflection of an object in a plane mirror is a virtual image the same size as the object.

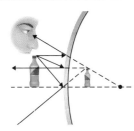

The image formed by a curved mirror might have a different size and might be inverted.

FOLDABLES

Use your lesson Foldable to review the lesson. Save your Foldable for the project at the end of the chapter.

What do you think NOW?

You first read the statements below at the beginning of the chapter.

3. Mirrors are the only surfaces that reflect light.

4. The image in a mirror is always right side up.

Did you change your mind about whether you agree or disagree with the statements? Rewrite any false statements to make them true.

Use Vocabulary

1 Distinguish between the focal point and the focal length of a mirror.

2 The reflection in a clear window of a store is a(n) _____.

3 A mirror that converges light rays is a _____ mirror.

Understand Key Concepts

4 Which term describes reflection from a rough surface?
   A. diffuse   C. regular
   B. irregular   D. translucent

5 Which terms describe the image of an object located between a concave mirror and its focal point?
   A. real, inverted   C. virtual, inverted
   B. real, upright   D. virtual, upright

6 Contrast your reflection from the back of a shiny spoon and from a plane mirror.

7 Explain why the image in a plane mirror is a virtual image.

Interpret Graphics

8 Organize Information Copy and fill in the graphic organizer below to describe how real and virtual images are produced by different types of mirrors.

Type of Mirror	How Images Form

Critical Thinking

9 Evaluate The rearview mirror on a car is often slightly convex. Explain why this is more practical than a plane mirror.

10 Construct a diagram showing the image that forms when an object is inside the focal point of a concave mirror.

Lesson 2 · **647**
EVALUATE

Inquiry Skill Practice — Measure

40 minutes

How can you demonstrate the law of reflection?

When light strikes a mirror, the angle of reflection equals the angle of incidence. You can demonstrate the law of reflection by measuring the angles of incident and reflected light rays between an object and a mirror.

Materials

modeling clay

plane mirror

ruler

protractor

Also needed:
paper, corrugated cardboard, pin

Safety

Learn It

One method of collecting data is to use a scientific tool and **measure** a value. You can use a protractor to measure the angles at which light strikes and reflects off a surface. You can use a ruler to measure distances of an object and its image from a mirror.

Try It

1. Read and complete a lab safety form.

2. Draw a line across a piece of paper. Label the line *Mirror*. Place the paper on cardboard so it will support pins.

3. Use clay to hold a mirror upright along the line, as shown in the picture.

4. Push a pin into the paper 5 cm in front of the mirror. Label this point on the paper *Object*.

5. With your eyes to the right of the pin, look toward the pin's image in the mirror. Draw a line on the paper from you to the image.

6. Have your partner repeat step 5 by looking at eye-level from the left of the pin.

7. Remove the pin and the mirror.

Apply It

8. First determine the location of the pin's virtual image. Extend the lines that you and your partner drew so that they meet at a point behind the mirror line. Label this point *Virtual image*.

9. Draw and label the incident ray from the object to the point where the line you drew in step 5 crossed the mirror. Draw a normal from the mirror at this point.

10. Light traveled from the object, toward this point, and back to your eye. Label the reflected ray.

11. Have your partner repeat steps 9–10 for the line drawn in step 6.

12. Draw your setup in your Science Journal. Measure and record the distance of the virtual image to the mirror.

13. Use a protractor to measure the angles of incidence and reflection for the light rays you and your partner saw. Record your measurements.

14. **Key Concept** How did your measurements support what you have learned about virtual images formed by a plane mirror? How did they demonstrate the law of reflection?

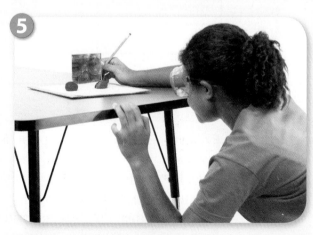

Lesson 3

Refraction and Lenses

Reading Guide

Key Concepts
ESSENTIAL QUESTIONS

- What happens to light as it moves from one transparent substance to another?
- How do convex lenses and concave lenses affect light?
- How do eyes detect light and color?

Vocabulary

refraction p. 650
lens p. 652
convex lens p. 652
concave lens p. 652
rod p. 657
cone p. 657

 Multilingual eGlossary

 Video BrainPOP®

Inquiry Why is its head so big?

The cat's head might look gigantic compared to its body, but it is just an illusion. Light can produce interesting effects when it moves from one transparent material, such as glass, to another transparent material, such as water. Why do you think this happens?

Inquiry Launch Lab

15 minutes

What happens to light that passes from one transparent substance to another?

Objects look different in water than in air. Light that reflects off objects changes when it moves from one transparent material to another.

1. Read and complete a lab safety form.
2. Pour about 150 mL of water into a **250-mL beaker.** Pour 150 mL of **vegetable oil** into a **second beaker.**
3. Place three **test tubes** in a **test-tube rack.** Leave one test tube empty. Half fill one with water. Half fill another with vegetable oil.
4. One at a time, place each test tube into the water beaker. Look through the side of the beaker at the test tube. Record your observations in your Science Journal.
5. Remove and dry the test tubes with a **paper towel.**
6. Repeat steps 4–5 using the oil beaker. Record your observations.

Think About This

1. Draw a diagram showing the substances light passed through as it moved from the test tube to your eye for each setup. Remember to include the glass of the test tube and the beaker.
2. **Key Concept** Summarize your observations of the test tubes of air, water, and oil.

Table 1 The index of refraction indicates how much a medium can change the direction of light.

Table 1

Medium	Index of Refraction
Vacuum	1.0000
Air	1.0003
Ice	1.31
Water	1.333
Oil	1.47
Ovenproof glass	1.47
Diamond	2.417

Refraction of Light

Have you ever tried to pick up something from the bottom of a container of water and the object was deeper than you thought it was? This happens because light waves can change direction. Light always travels through empty space at the same speed—300,000 km/s. Light travels more slowly through a medium (plural, media), such as air, glass, or water. The atoms of the material interact with the light waves and slow them down. Some substances, such as air, only slow light a little. Others, such as glass, slow the light more.

As a light wave moves from one medium into another, its speed changes. If it enters the new medium at an angle, the wave will change direction. *The change in direction of a wave as it changes speed while moving from one medium to another is called* **refraction.**

Each transparent material has a property called the index of refraction, as shown in **Table 1.** A medium that has a high index of refraction is sometimes called slow because light moves more slowly through it. A medium that has a relatively low index of refraction, such as air, is called fast.

Key Concept Check What happens to light as it moves between transparent substances?

Moving Into a Slower Medium

Suppose you roll a toy car across a table straight at a piece of fabric. The front tires of the car slow down when they hit the fabric. The car continues to move in a straight line but more slowly. If you roll the car at an angle, one of the front tires will hit the fabric before the other. That side of the car will slow down, but the rest of the car will continue at the same speed until the other tire hits the fabric. This will cause the car to turn and change direction.

As shown in **Figure 14,** a light wave behaves in a similar way when it moves into a slower medium. Recall that a normal is a line perpendicular to a surface. As light moves into a slower medium at an angle, it changes direction toward the normal.

Moving Into a Faster Medium

What happens when light in the figure moves back into the air? Suppose you ride your bike from a sidewalk into a muddy field and then back onto a sidewalk. You use the same energy to pedal the whole time, but you move more slowly in the mud. When you move back onto the sidewalk, you speed up.

Similarly, as light moves into a medium with a lower index of refraction, it speeds up. The wave is still at an angle, so the part that leaves the slower medium first, speeds up sooner. This causes the wave to turn away from the normal, as shown on the right in **Figure 14.**

You see the boundaries between surfaces such as air, glass, and water because of refraction. If transparent substances have the same index of refraction, you do not see the surfaces. The Launch Lab at the start of this lesson shows this effect.

Math Skills

Use Scientific Notation

You can calculate the index of refraction of a material by dividing the speed of light in a vacuum (3.0×10^8 m/s) by the speed of light in that material. When dividing numbers written in scientific notation, divide the coefficients and subtract the exponents. For example, what is the index of refraction of a substance in which light travels at 2.0×10^8 m/s?

a. Set up the problem.

$$\text{Index of refraction} = \frac{3.0 \times 10^8 \text{ m/s}}{2.0 \times 10^8 \text{ m/s}}$$

Note that units cancel.

b. Divide the coefficients.

$$3.0 \div 2.0 = 1.5$$

c. Subtract the exponents.

$$8 - 8 = 0$$

The index of refraction is $1.5 \times 10^0 = 1.5 \times 1 = 1.5$.

Practice

The speed of light in a material is 1.56×10^8 m/s. What is its index of refraction?

- Review
- Math Practice
- Personal Tutor

Refraction

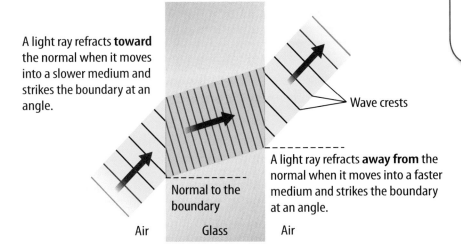

A light ray refracts **toward** the normal when it moves into a slower medium and strikes the boundary at an angle.

Wave crests

A light ray refracts **away from** the normal when it moves into a faster medium and strikes the boundary at an angle.

Normal to the boundary

Air Glass Air

Figure 14 The direction of refraction depends on each material's index of refraction.

Review Personal Tutor

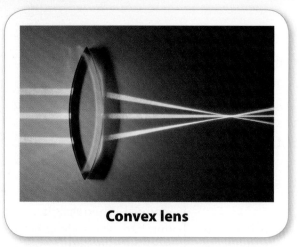
Convex lens

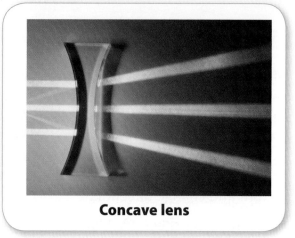
Concave lens

Figure 15 A convex lens often is called a converging lens. A concave lens often is called a diverging lens.

Lenses

What do binoculars, eyeglasses, your eyes, and a camera have in common? Each contains a lens. *A **lens** is a transparent object with at least one curved side that causes light to change direction.* Recall that most of the light that strikes a transparent material passes through it. Light refracts as it passes through a lens. The greater the curve of the lens, the more the light refracts. The direction of refraction depends on whether the lens is curved outward or inward.

As shown in **Figure 15,** light passes through two different lenses. *A lens that is thicker in the middle than at the edges is a **convex lens.*** Notice that the light rays that move through a convex lens come together, or converge. *A lens that is thicker at the edges than in the middle is a **concave lens.*** Notice that the light rays spread apart, or diverge, as they move through a concave lens.

Reading Check Describe the shapes of a convex lens and a concave lens.

Inquiry MiniLab
15 minutes

How can water move light?
You can draw light rays to investigate refraction of light.

1. Read and complete a lab safety form.
2. Place a **sheet of paper** over **corrugated cardboard.** Place a **clear plastic box** on the paper. Use a pencil to trace the long sides of the box. Fill the box with water.
3. Place two **pins** into the paper and cardboard on opposite edges of the box. With one eye, look from one pin to the other, diagonally through the box. Imagine a line through the pins. Draw along this line extending out from each pin and away from the box.
4. Remove the box. Use a **ruler** to draw a line between the pins.

Analyze and Conclude
1. **Draw** the light's path as it moved through the box to your eye. Label each medium the light passed through.
2. **Key Concept** Identify and explain each change in the light's direction as it moved among the different media.

Convex Lenses

Have you ever used a magnifying lens to look at a tiny insect? A magnifying lens makes things appear larger. If you look closely, you can see that a magnifying lens is a convex lens because its center is thicker than its edges.

The refraction of light by a convex lens is shown in **Figure 16**. Notice that a normal to the curved surface slants toward the optical axis. Recall that light moving into a slower *medium* turns toward the normal. As a result, a convex lens refracts light inward and it converges.

Focal Point and Focal Length Similar to a mirror, the point where rays parallel to the optical axis converge after passing through a lens is the focal point. The distance along the optical axis between the lens and the focal point is the focal length of the lens. Because you can look through a lens from either side, a focal point is on both sides of the lens. For a lens with the same curve on both sides, the lens's two focal points are the same distance from it.

 Key Concept Check What happens to light as it passes through a convex lens?

Types of Images Like a concave mirror, the type of image a convex lens forms depends on the location of the object. A convex lens can form both real and virtual images, as shown in **Figure 17**. The diagrams show only two rays, but remember that in reality there are an infinite number of rays.

If you look through a magnifying lens at an object more than one focal length from the lens, the image you see is inverted and smaller. If you look at an object less than one focal length from the lens, the image you see is upright and larger. It is virtual because your brain interprets the rays as moving in a straight line, as shown by the dashed lines in **Figure 17**.

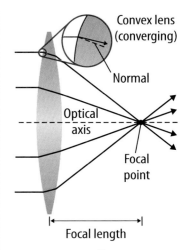

▲ **Figure 16** The outward curve of a convex lens refracts light rays inward.

SCIENCE USE V. COMMON USE

medium
Science Use matter through which a wave travels

Common Use middle or average

Figure 17 You can see either a virtual image or a real image if you look through a convex lens. ▼

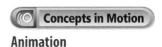
Animation

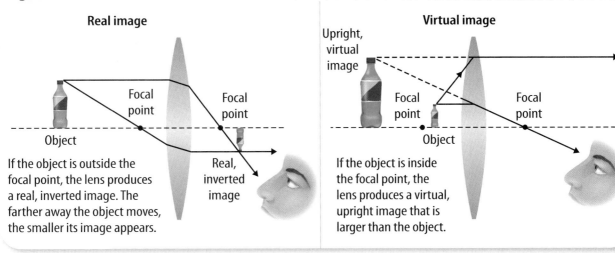

If the object is outside the focal point, the lens produces a real, inverted image. The farther away the object moves, the smaller its image appears.

If the object is inside the focal point, the lens produces a virtual, upright image that is larger than the object.

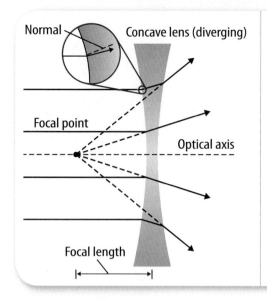
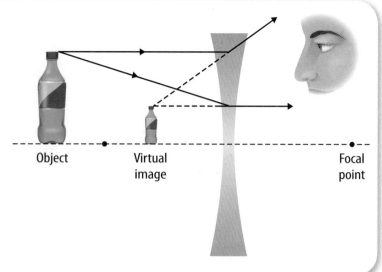

▲ **Figure 18** 🔑 The inward curve of a concave lens refracts light rays outward. You see a virtual image if you look through a concave lens.

Concave Lenses

Light rays that pass through a concave lens diverge, or spread apart. The drawing on the left in **Figure 18** shows why. Notice that a normal to the curved surface slants away from the optical axis. Because light entering a slower medium changes direction toward the normal, the lens refracts light outward. The light diverges.

The drawing on the right in **Figure 18** shows the type of image a concave lens forms. Suppose you stand to the right of the lens and look at the object. The refracted rays reach your eyes, but your brain assumes the light traveled the straight dashed lines. You see a virtual image smaller than the object.

 Key Concept Check What happens to light as it passes through a concave lens?

Refraction and Wavelength

You have read that the curved surfaces of lenses affect the refraction of light rays. The wavelength of light also affects the amount of refraction. You might have seen sparkling cut-glass ornaments hanging in a window. When sunlight strikes the glass, rainbows appear on the wall. These "sun catchers" work because different wavelengths of light refract different amounts.

You read in Lesson 1 that white light is made up of different colors. A few colors are shown in **Figure 19.** The speed of a wave in a material is related to its wavelength. Waves with longer wavelengths travel at greater speeds in a material than waves with shorter wavelengths. Therefore, when entering a material, light with longer wavelengths travels faster and refracts less than light with shorter wavelengths. As a result, violet refracts the most because its wavelength is the smallest. Red has the longest wavelength and refracts the least.

Figure 19 Each color of light has a different wavelength and frequency. ▼

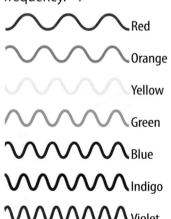

◀ **Figure 20** A prism spreads light into different wavelengths.

Visual Check Compare the refraction of yellow and blue light by a prism.

Prisms

As white light passes through a piece of glass called a prism, each wavelength of light refracts when it enters and again when it leaves. Because each color refracts at a slightly different angle, the colors separate. They spread out into the familiar spectrum, as shown in **Figure 20.**

Rainbows

Why do rainbows appear in the sky only during or after a rain shower? Rainbows form when water droplets in the air refract light like a prism. Each wavelength of light refracts as it enters the droplet, reflects back into the droplet, and refracts again when it leaves the droplet. Notice in **Figure 21** that wavelengths of light near the blue end of the spectrum refract more than wavelengths near the red end of the spectrum. This effect produces the separate colors you see in a rainbow.

FOLDABLES

Make a vertical two-tab book. Label it as shown. Use it to summarize your notes on prisms and rainbows.

Figure 21 Sunlight refracts as it passes into and out of a raindrop. ▼

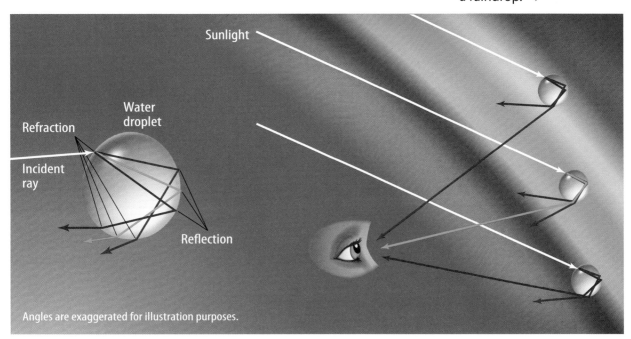

Visual Check Why do you have to stand with the Sun behind you to see a rainbow?

Detecting Light

You receive much information about the world around you when objects either emit light or reflect light into your eyes. You have read that the material that makes up an object reacts to various wavelengths of light in different ways. The material absorbs some wavelengths of light. Other wavelengths of light reflect to your eyes, providing information about shapes and colors. Your brain interprets this light energy so that you can recognize the images you see as people, places, and objects.

The Human Eye

Eyes respond quickly to changing conditions. As shown in **Figure 22,** the iris changes the size of the pupil, which controls the amount of light that enters an eye. The cornea acts as a convex lens and focuses light when it enters the eye. The shape of the cornea cannot change, but the shape of the eye's lens can. Changes to the shape of the lens can alter its focal point. This can enable a person to see objects clearly either far away or near. Follow the steps in the **Figure 22** to learn what each part of the eye does that enables a person to see.

How the Eye Works 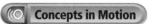 Concepts in Motion Animation

Figure 22 The parts of an eye work together to focus light and send signals about what you see to the brain.

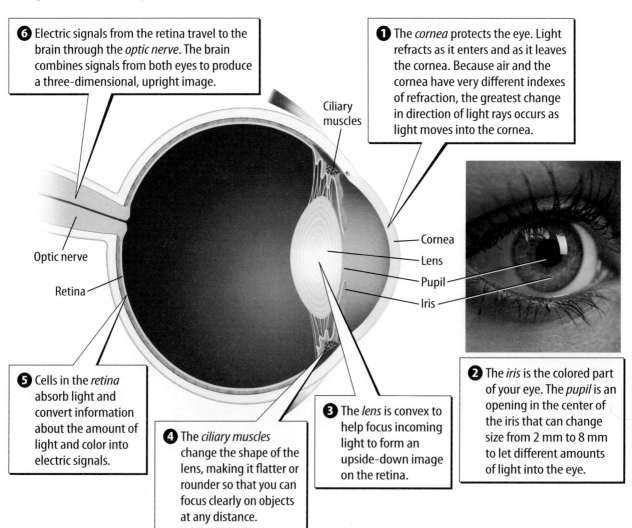

Seeing Color

You might have noticed that it is harder to see colors in dim light. Why does this happen? Different cells on the retina of an eye enable a person to see both colors and shades of gray. *A **rod** is one of many cells in the retina of the eye that responds to low light.* There are about 120 million rods in the human retina. Rods enable you to see near the sides of your eyes rather than along the direct line of vision. This type of eyesight is called peripheral [puh RIH fuh rul] vision.

*A **cone** is one of many cells in the retina of the eye that respond to colors.* The human retina contains 6 million to 7 million cones that require more light to produce signals. These cones respond to red, green, and blue wavelengths of light. Recall that these primary colors of light can combine and form all the other colors. When light strikes a cone, it produces an electric signal that depends on the light's intensity and its wavelength. Your brain combines signals from all rods and cones and forms the colors, the shapes, and the brightness of the objects you see, as shown in **Figure 23**.

▲ **Figure 23** The ability to distinguish colors allows you to see more detail in your surroundings.

WORD ORIGIN

cone
from Latin *conus*, means "wedge"

 Key Concept Check How do eyes detect light and color?

Correcting Vision

In a person with normal eyesight, the cornea and the lens focus light directly on the retina. This forms a sharp, clear image. Some eyes, however, have an irregular shape. If an eye is too long, light focuses in front of the retina. This person sees near objects clearly, but objects far away are blurred. In other instances, the eye can be too short, and light focuses behind the retina. People with this problem clearly see objects far away, but objects nearby are blurred. **Figure 24** shows how manufactured lenses can correct these problems.

You have read that refraction and lenses enable you to see. Light can refract as it moves from one medium into another. Different colors refract at different angles. A lens causes light to change direction. An example is the focusing of light onto the retina by the cornea and lens of the eye. Eyeglasses or contact lenses may be needed to help focus this light.

Figure 24 Some people have eyes with irregular shapes that require lenses to correct their vision. ▼

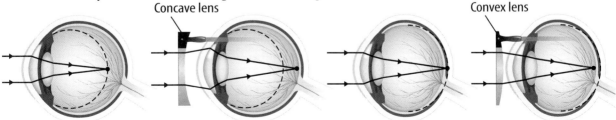

A person is **nearsighted** when light focuses in front of the retina. A **concave** lens corrects the problem.

A person is **farsighted** when light focuses behind the retina. A **convex** lens corrects the problem.

Lesson 3 Review

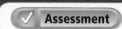 Assessment Online Quiz

Visual Summary

Refraction is the change in direction of a wave as it moves from one medium into another at an angle.

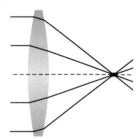

A lens changes the direction of light rays that pass through it.

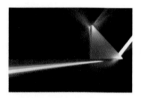

Each color of light refracts differently as it passes through a prism.

FOLDABLES

Use your lesson Foldable to review the lesson. Save your Foldable for the project at the end of the chapter.

What do you think NOW?

You first read the statements below at the beginning of the chapter.

5. A prism adds color to light.

6. When light moves from air into water or glass, it travels more slowly.

Did you change your mind about whether you agree or disagree with the statements? Rewrite any false statements to make them true.

Use Vocabulary

1 **Identify** two differences in rods and cones.

2 **Describe** a convex lens in your own words.

Understand Key Concepts

3 Which terms describe the image produced by a concave lens?
- A. real, inverted
- B. real, upright
- C. virtual, inverted
- D. virtual, upright

4 **Contrast** refraction when light moves into an area of higher index of refraction and into an area of lower index of refraction.

5 **Identify** the functions of the pupil, the cornea, the lens, and the retina.

Interpret Graphics

6 **Examine** What does the direction of the light wave tell you about the index of refraction of each of the materials below?

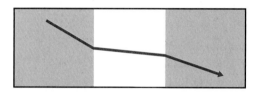

7 **Sequence** Copy and fill in a graphic organizer like the one below that shows what happens as light enters and interacts with parts of the eye and allows you to see an image. Add boxes as necessary.

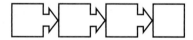

Critical Thinking

8 **Predict** How would an object appear if you looked at it through a concave lens and a convex lens at the same time? Why?

Math Skills

Review — Math Practice

9 Light travels through a sheet of plastic at 1.8×10^8 m/s. What is the index of refraction of the plastic?

Inquiry Skill Practice
Gather and Organize Information — 40 minutes

How does a lens affect light?

Materials

index card

convex lens

ruler

meterstick

Also needed:
concave lens

Safety

Light changes direction as it passes through a lens. This can make an object appear larger or smaller. It can even make the object appear upside down. The shape of the lens and the distance of the lens from the object affect the image you see.

Learn It
Making observations and measuring are two ways to **gather information** during an investigation. You can **organize information** you gather by recording it in a data table.

Try It

1. Read and complete a lab safety form.

2. Copy the table below into your Science Journal. Add lines as needed.

3. Prop a textbook upright. Hold a convex lens at arm's length and 1.5 m from the book. Look at the book through the lens. Record your observations.

4. Observe the image as you slowly walk toward the book, keeping the lens at arm's length. Have your partner measure the distance from the lens to the book whenever the image changes significantly. Record your data and observations.

5. With the classroom lights dimmed, hold a card in front of a window. Hold the lens between the card and the window, as shown.

6. Try to focus an image of the scene outside onto the card by moving the card closer and farther from the lens. When the image is focused as clearly as possible, have your partner measure the distance between the lens and the card. Record data and observations.

7. Repeat steps 3–6 with a concave lens.

Apply It

8. **Identify** positions of each lens when you saw significant changes in the image. What caused these?

9. **Infer** the focal length of the convex lens.

10. 🔑 **Key Concept** Summarize your observations of the convex-lens and concave-lens images.

Type of Lens	Distance from the Lens to the Object	Description of the Object
Convex		

Lesson 3
EXTEND
659

Lesson 4

Reading Guide

Key Concepts
ESSENTIAL QUESTIONS

- What do devices such as telescopes, microscopes, and cameras have in common?
- What is laser light, and how is it used?
- How do optical fibers work, and how are they used?

Vocabulary
optical device p. 661
refracting telescope p. 662
reflecting telescope p. 662
microscope p. 662
laser p. 664
hologram p. 665

 Multilingual eGlossary

 Video
What's Science Got to do With It?

Optical Technology

Inquiry How does the light do that?

A laser-light show is a spectacular site as the beams of light streak through the air. The lights in your home and school fill a room with brightness, but laser lights do not. How do laser lights travel in such a straight line? What makes them different from other lights?

Inquiry Launch Lab

10 minutes

How do long-distance phone lines work?

Long-distance telephone calls often are sent as light signals along thin strands of glass called optical fibers. Why doesn't light escape out the sides of the fibers?

1. Read and complete a lab safety form.
2. Fill a **clear plastic box** with water.
3. In a darkened room, shine a thin **flashlight** beam through the side of the box into the water. Stir **milk** into the water one drop at a time until you can see the beam in the water.
4. Position the beam of light through the side of the box so that it is almost perpendicular to the underside of the water's surface. Record your observations of the reflected beam and the water in your Science Journal.
5. Hold an **index card** above the box so that any light transmitted through the surface shines on it.
6. Slowly change the angle of the beam until no light is transmitted onto the card. Draw your setup, and record your observations in your Science Journal.

Think About This

1. How did the angle of the incident beam affect the amount of light that appeared on the card?
2. **Key Concept** How do you think your observations show what happens to light in an optical fiber?

Optical Devices That Magnify

Do you wear sunglasses or contact lenses? Have you ever watched an animal through binoculars or looked at a planet through a telescope? If you answered yes to any of these questions, you have used an optical device. *Any instrument or object used to produce or control light is an* **optical device.** Some optical devices use lenses to help you see small or distant objects.

Telescopes

Hundreds of years ago, people could not see details of the Moon, such as those shown in **Figure 25**. However, by using telescopes, people have learned about objects in space. Telescopes are optical devices that produce magnified images of distant objects. A telescope uses a lens or a concave mirror to gather light. You might have used a small telescope to observe the Moon or planets, but some telescopes are much larger. The largest optical telescope in the United States has a mirror that is 10 m across. This enormous size means that the telescope can gather more light from a distant object than your eye can. The *Hubble Space Telescope* orbits about 600 km above Earth. It has a clearer view because it is above Earth's atmosphere. Astronomers use it to see objects billions of light-years away.

Figure 25 A telescope gives a close-up view of the Moon's surface.

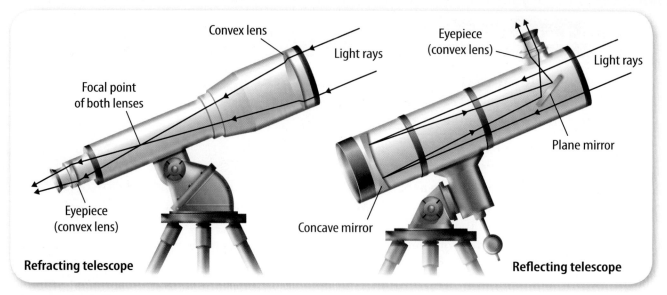

▲ **Figure 26** Telescopes use either a lens or a mirror to gather light from a faraway source.

WORD ORIGIN

microscope
from Latin *microscopium*, means "an instrument used to view small things"

Figure 27 The main difference in microscopes and refracting telescopes is focal length. ▼

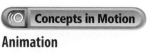

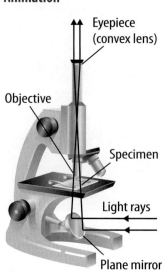

Refracting Telescope *A telescope that uses lenses to gather and focus light from distant objects is a* **refracting telescope.** The convex lens that gathers light is called the objective. As shown on the left in **Figure 26,** light rays that pass through the objective converge and form a real image. The convex lens that a viewer looks through is called the eyepiece. This lens magnifies the image. Recall that refraction differs slightly for different wavelengths of light. Eyepieces often include multiple lenses with different focal lengths so that all wavelengths of light focus clearly.

Reflecting Telescope *A telescope that uses a mirror to gather and focus light from distant objects is a* **reflecting telescope.** A simple reflecting telescope is shown on the right in **Figure 26.** Light enters the tube and reflects off a concave mirror at the far end. The light begins to converge, but before it converges completely, it reflects off a plane mirror toward the eyepiece. A real image forms, and the eyepiece lens magnifies the image.

Light Microscopes

Telescopes are useful for observing far away objects, but to observe tiny objects up close, you need a different instrument. *A* **microscope** *is an optical device that forms magnified images of very small, close objects.* A microscope works like a refracting telescope, but its lenses have shorter focal lengths. As shown in **Figure 27,** light from a mirror or a light source near the microscope's base passes through a thin sample of an organism or object. The light then travels through a convex lens, called an objective, that focuses it. The eyepiece magnifies the image. Microscopes often have several objective lenses with different focal lengths for different magnifications.

Key Concept Check How are refracting telescopes and microscopes alike?

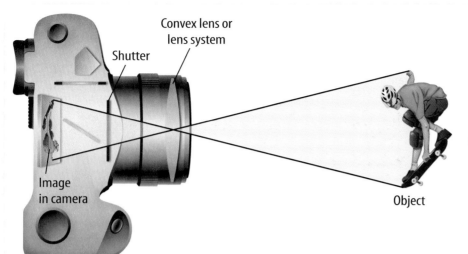

Figure 28 🔑 A digital camera uses one or more convex lenses to focus light onto a sensor.

✓ **Visual Check** Why would the lens system have to move in order to photograph more distant objects?

Cameras

Cameras come in many different designs that use lenses to focus images. The design of a typical digital camera is illustrated in **Figure 28.** Suppose you want to take a photo of your friend on a skateboard. Light reflects off your friend and enters the camera. The light passes through a convex lens or, more often, a lens system. A shutter normally blocks light from entering the camera. However, pressing a button briefly opens the shutter. Light enters and is detected by an electronic sensor in the back of the camera. Recall that a convex lens is a converging lens. Inside the camera, a lens or system of lenses converges the light rays toward the sensor.

The boy you photograph is farther away from the lens than the focal point. For a convex lens, this means the image produced is real, inverted, and smaller than the object. Notice in the figure that this is the type of image that appears on the electronic sensor.

With some cameras, you can zoom, or move the lenses and increase the size of an image. Zooming changes the lenses' focal lengths. This means a camera can take enlarged images of distant or close objects.

✓ **Key Concept Check** What does a camera have in common with a telescope and a microscope?

Inquiry MiniLab
15 minutes

How does a zoom lens work?

A zoom lens can focus on distant objects. You can use a convex lens and a concave lens to observe the basic principle behind how zoom lenses work.

1. Read and complete a lab safety form.
2. Hold a **convex lens** at arm's length. Look through the lens at a distant object. Record your detailed observations in your Science Journal.
3. Hold a **concave lens** between you and the convex lens. Look through both lenses at the same object. Record your observations. Note changes in the image.
4. Move the concave lens closer and then farther from the convex lens. Does the image change? Record your observations.

Analyze and Conclude

1. **Contrast** the image with the convex lens with the image using both lenses.
2. 🔑 **Key Concept** Based on your observations, what do you think is one reason telescopes, microscopes, and cameras often have systems of lenses for focusing rather than just one lens?

Lesson 4
EXPLAIN

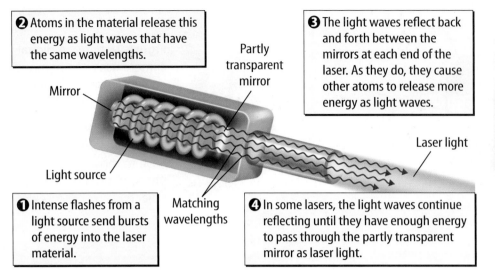

❷ Atoms in the material release this energy as light waves that have the same wavelengths.

❸ The light waves reflect back and forth between the mirrors at each end of the laser. As they do, they cause other atoms to release more energy as light waves.

❶ Intense flashes from a light source send bursts of energy into the laser material.

❹ In some lasers, the light waves continue reflecting until they have enough energy to pass through the partly transparent mirror as laser light.

▲ **Figure 29** Light from the laser spreads much less than light from other sources because the light waves have the same wavelength.

Figure 30 Most DVDs hold over ten times more information than a CD. For example, you can record an entire movie on one DVD. ▼

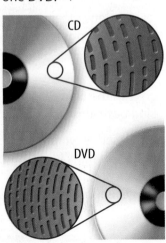

Lasers

Some lasers are gentle enough to operate on the human eye, but some are powerful enough to cut through steel. *A **laser** is an optical device that produces a narrow beam of coherent light.* Recall that light from a luminous source, such as a lightbulb, has many wavelengths and spreads out in all directions. Coherent light is different. The light waves of coherent light all have the same wavelength and travel together. The beam is narrow because the waves do not spread apart, unlike light from other sources. Also, the intensity of laser light does not decrease quickly as it moves away from its source.

The word *laser* stands for light amplification through stimulated emission of radiation. One type of laser is shown in **Figure 29.** An energy source causes atoms of a material in the laser to emit light. This is the stimulated emission. This light travels back and forth within a tube, causing other atoms to emit similar waves. In this way, the light intensity increases. This is the light amplification.

 Key Concept Check What is laser light?

CDs and DVDs

Whenever you use a CD player, you use a laser. To make a CD, electric signals represent sounds. A laser burns tiny pits in the surface of the CD that correspond to the signals. In your CD player, another laser passes over the pits as the CD spins. Different amounts of light reflect back to a sensor that converts the light back into an electric signal. The signal causes the speakers to produce sound.

DVDs are similar to CDs except the DVD's pits are much smaller and closer. Therefore, more information fits on a DVD than a CD, as shown in **Figure 30.**

Holograms

Perhaps you've seen images on a book cover or on a credit card that seem to float in space. You can see different sides of the object, just as you would with a real object. A **hologram** is *a three-dimensional photograph of an object.*

Figure 31 shows how one type of hologram forms. First, laser light splits into two beams. One beam passes through a convex lens then reflects from the object onto photographic film. The other beam passes through a convex lens and then travels to the film. The combined light from the two beams produces a pattern on the film that shows both the brightness of light and its direction. The pattern appears as swirls, but when a laser shines on the film, a holographic image appears.

Some paper bills and credit cards have a different type of hologram printed on them to prevent counterfeiting. The holograms are made with lasers but can be viewed under regular light. Other uses of lasers are shown in **Figure 32**.

 Key Concept Check What are some ways lasers are used?

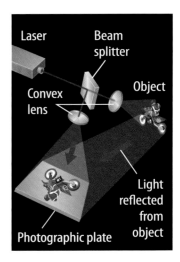

▲ **Figure 31** Interference of two laser-light beams produces a hologram.

Figure 32 Laser light is useful because its intensity does not decrease quickly. ▼

▲ Doctors use low-power lasers to correct problems in the retina.

▲ Lasers scan barcodes at stores and libraries.

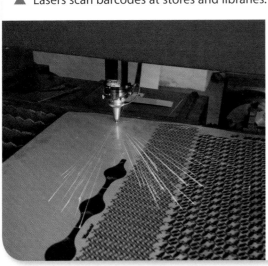

▲ Lasers guide tunnel-boring machines to keep tunnels straight.

◀ Powerful lasers cut through metal to form machine parts or intricate designs.

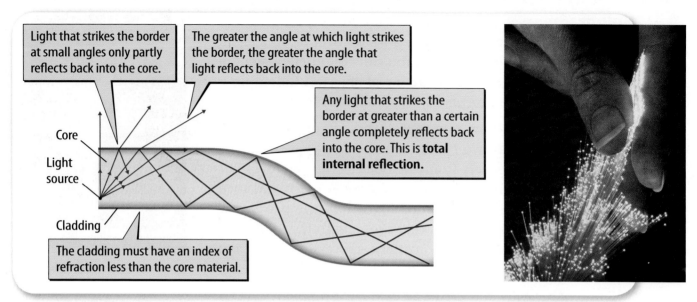

Figure 33 Light moves through an optical fiber because of a phenomenon called total internal reflection.

Visual Check Why does light stay inside an optical fiber?

Optical Fibers

Almost every time you make a phone call, use a computer, or watch a television show, you are using light technology. The electronic signals for these devices travel at least part of the way through optical fibers. An optical fiber is a thin strand of glass or plastic that can transmit a beam of light over long distances.

Total Internal Reflection Just as water moves through a pipe, light moves through an optical fiber. **Figure 33** shows how this happens. A core material surrounded by another material called the cladding makes up an optical fiber. As light moves through the core, it strikes the cladding at an angle greater than a certain angle and reflects back into the core. This process is called total internal reflection. Because light stays in the core, a fiber can carry light signals long distances without losing strength.

Uses of Optical Fibers Phone conversations, computer data, and TV signals travel as pulses of light through optical fibers. Using different wavelengths of light, one fiber can carry thousands of phone conversations at a time. Optical fibers also have medical uses, such as providing light for a physician to perform surgery through a small incision in the body.

Key Concept Check How do optical fibers work, and how are they used?

Relating Optical Technology Each of the optical devices that you have read about uses lenses or mirrors to produce or control light. Telescopes and light microscopes produce magnified images. Cameras use lenses to focus light on an image sensor. Lasers produce coherent light. Optical fibers are also a type of optical device because they transmit and reflect light.

FOLDABLES
Use two sheets of paper to make a bound book and label it as shown. Use it to organize what you have learned about different types of optical devices.

Lesson 4 Review

Assessment | **Online Quiz**

Visual Summary

A telescope gathers and focuses light and produce a magnified image of an object far away.

Lasers have many uses because they produce an intense beam of light.

Almost no light is lost as light passes through an optical fiber.

FOLDABLES

Use your lesson Foldable to review the lesson. Save your Foldable for the project at the end of the chapter.

What do you think NOW?

You first read the statements below at the beginning of the chapter.

7. A laser will burn a hole through your skin.

8. Telephone conversations can travel long distances as light waves.

Did you change your mind about whether you agree or disagree with the statements? Rewrite any false statements to make them true.

Use Vocabulary

1 Use the term *microscope* in a sentence.

2 Define *hologram* in your own words.

3 An optical device that produces a narrow beam of coherent light is a(n) _____.

Understand Key Concepts

4 Which collects light in a reflecting telescope?
 A. concave lens C. convex lens
 B. concave mirror D. convex mirror

5 **Contrast** a microscope and a camera.

6 **Describe** how coherent light is produced in a laser.

7 **Explain** total internal reflection.

Interpret Graphics

8 **Identify** the optical effect that a jeweler achieves when he cuts facets, or faces, on a diamond, as shown here. Why does cutting increase the sparkle?

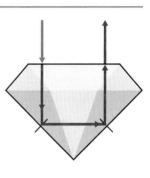

9 **Compare and Contrast** Copy and fill in a graphic organizer like the one below. How are the lenses in a refracting telescope and a microscope alike and different?

Critical Thinking

10 **Compare and contrast** a hologram and a regular image taken by a camera.

11 **Extend** When you look through binoculars, each eye looks through two convex lenses. Which other optical devices work in this way?

Inquiry Lab

3 class periods

Design an Optical Illusion

Materials

beakers

clay

color filters

flashlights

markers

prisms

Also needed:
various lenses, various mirrors, cardboard box, plastic box, scissors, stands, candles, matches, test tubes, water, vegetable oil

Safety

A famous illusionist has hired you to design a new type of optical illusion for a show. How will you do it? To make an optical illusion, you have to use lights to make people see an image that is not really there. Think about how you can use all you have learned about light and optical devices to make an illusion. Here is your challenge:

- You may use only common materials and devices approved by your teacher.
- You must use light in at least two different ways (transmit, reflect, refract, or absorb).
- The image you create must be different from the object(s) in at least two ways (for example, size, orientation, color, or position).

Question

How can you design and build an optical illusion? How can light be used to create an optical illusion? What materials are easily manipulated to construct a simple optical illusion?

Procedure

1. Read and complete a lab safety form.
2. Brainstorm ideas for your illusion with others in your group. Think about light sources and shadows. Can you combine images or show only part of an image? How does each optical device work, and how could you combine them? Record your ideas in your Science Journal.
3. Use your ideas to design an optical illusion. List the materials and steps you will follow to construct it. Include sketches of your setup.
4. Have your teacher approve your design.
5. Collect your materials and build your optical illusion.
6. Test your illusion several times. Record your observations.

Analyze and Conclude

7 Evaluate the quality of your illusion. Consider these questions: *Does it meet all the requirements? Is your illusion creative and unique? How easy is it for an audience to see? How different is the image your illusion produces from the object?*

8 Make a modification to your setup. Test and evaluate the effectiveness of your setup in producing a realistic illusion.

9 Record details about your modification. Then record your observations.

10 Continue to improve your illusion as time permits.

11 Sketch and label your final setup. Indicate the path of light.

12 Critique your design and results. Did your setup work the way you expected? Did it meet all the requirements?

13 The Big Idea How did matter and light interact to produce your illusion?

Lab Tips

☑ Think about how you might combine images of objects that are far apart into a single image.

☑ Consider how you could use specific colors in your illusion or how you could use filters to block some colors of light.

☑ Remember that you can use lenses and mirrors to make images appear in different locations than where the objects are.

Communicate Your Results

Demonstrate your illusion to the class. Imagine that you are presenting the illusion to the illusionist who hired you. Explain your design. The illusionist needs to understand how all the parts work in order to present the illusion to an audience. Describe how light travels between media. Explain how reflection and refraction affect the path of the light to produce an illusion.

Inquiry Extension

Explain how you could modify your illusion for use on stage. This might include making the setup larger so that an entire theater audience could see the illusion. Also consider whether you would be able to take your setup apart and put it together again as the show travels from one city to another.

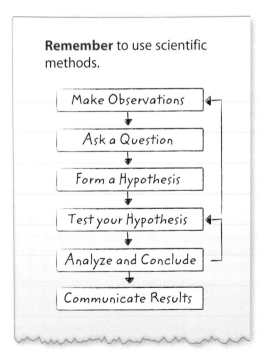

Remember to use scientific methods.

Make Observations → Ask a Question → Form a Hypothesis → Test your Hypothesis → Analyze and Conclude → Communicate Results

Chapter 18 Study Guide

THE BIG IDEA: Matter can absorb, reflect, or transmit light. Objects might appear as different colors or sizes because of the way light interacts with matter.

Key Concepts Summary

Lesson 1: Light, Matter, and Color

- Luminous objects, such as a flashlight, are sources that produce **light** that spreads out in straight lines.
- Matter can emit, absorb, reflect, or transmit light.
- An object's color is determined by the wavelengths of light the object reflects or transmits to you.

Lesson 2: Reflection and Mirrors

- Light rays reflect parallel from smooth surfaces and in many directions from rough surfaces.
- The shape of a mirror affects the way it reflects light. **Concave mirrors** converge light rays. **Convex mirrors** diverge light rays.
- All mirrors can produce virtual images, but only concave mirrors can produce real images.

Lesson 3: Refraction and Lenses

- Light changes direction if it moves into a medium with a different index of **refraction**.
- **Convex lenses** converge light rays. **Concave lenses** diverge light rays.
- **Rods** in the retina of the eye detect the intensity of light. **Cones** detect the colors of objects.

Lesson 4: Optical Technology

- Telescopes, **microscopes**, and cameras use lenses to focus light.
- **Lasers** produce a narrow beam of coherent light. Uses of laser light include detecting information on DVDs, making **holograms**, and cutting metal.
- Light completely reflects back into optical fibers. These fibers can carry light signals over long distances.

Vocabulary

light p. 635
reflection p. 635
transparent p. 636
translucent p. 636
opaque p. 636
transmission p. 637
absorption p. 637

law of reflection p. 643
regular reflection p. 644
diffuse reflection p. 644
concave mirror p. 645
focal point p. 645
focal length p. 645
convex mirror p. 646

refraction p. 650
lens p. 652
convex lens p. 652
concave lens p. 652
rod p. 657
cone p. 657

optical device p. 661
refracting telescope p. 662
reflecting telescope p. 662
microscope p. 662
laser p. 664
hologram p. 665

Study Guide

- Personal Tutor
- Vocabulary eGames
- Vocabulary eFlashcards

FOLDABLES Chapter Project

Assemble your lesson Foldables as shown to make a Chapter Project. Use the project to review what you have learned in this chapter.

Use Vocabulary

1. You can see light but not objects clearly through a(n) _____ object.

2. Contrast opaque and transparent objects.

3. A mirror that causes light waves to converge is a(n) _____.

4. The place where rays parallel to the optical axis cross after reflecting from a mirror is called the _____.

5. Describe refraction in your own words.

6. A three-dimensional image of an object is a(n) _____.

7. Which optical instrument uses at least two convex lenses to magnify small, close objects?

Link Vocabulary and Key Concepts

 Interactive Concept Map

Copy this concept map, and then use vocabulary terms from the previous page to complete the concept map.

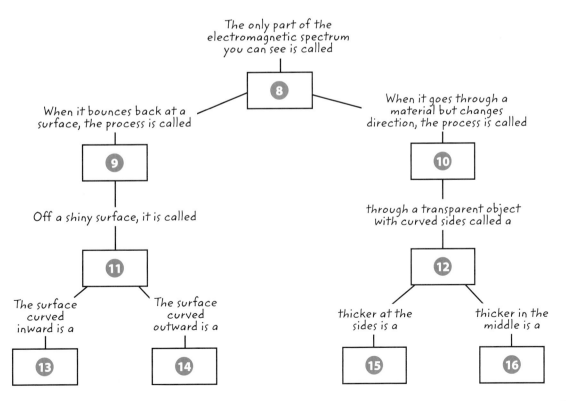

Chapter 18 Study Guide • **671**

Chapter 18 Review

Understand Key Concepts

1 Which best describes the image formed in a plane mirror?
   A. a real image behind the mirror
   B. a real image in front of the mirror
   C. a virtual image behind the mirror
   D. a virtual image in front of the mirror

2 Which MOST determines the color of an opaque object?
   A. diffraction
   B. reflection
   C. refraction
   D. transmission

3 Which describes a material that does NOT transmit any light?
   A. opaque
   B. reflective
   C. translucent
   D. transparent

4 If the angle of incidence of a ray striking a plane mirror is 50°, what is the angle of reflection?
   A. 40°
   B. 50°
   C. 100°
   D. 140°

5 Which terms describe the image that forms as light passes through the lens shown below?

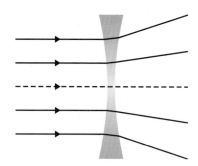

   A. real image; larger than the object
   B. real image; smaller than the object
   C. virtual image; larger than the object
   D. virtual image; smaller than the object

6 What color of light is produced when the three primary colors are combined in equal amounts?
   A. black
   B. blue
   C. red
   D. white

Use the diagram below to answer questions 7–9.

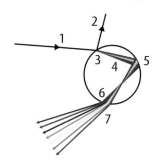

7 The diagram above shows light entering a raindrop. What produced the light represented by ray 2?
   A. It was absorbed by the drop.
   B. It was reflected from the drop.
   C. It was refracted by the drop.
   D. It was transmitted by the drop.

8 What caused the spread of the colors at point 7?
   A. absorption
   B. reflection
   C. refraction
   D. transmission

9 What process took place at position 5?
   A. absorption
   B. holography
   C. coherent light refraction
   D. total internal reflection

10 What term describes the degree to which a material causes light to move more slowly?
   A. focal length
   B. focal point
   C. index of refraction
   D. optical axis

672 • Chapter 18 Review

Chapter Review

Assessment
Online Test Practice

Critical Thinking

11 Analyze A girl wants to take a picture of her reflection in a plane mirror. She stands 1.3 m in front of the mirror. At what distance should she focus the camera to get a clear image of her reflection?

12 Hypothesize How might a magician make an object seem to disappear using a concave mirror?

13 Infer A solid, transparent object is placed in a container of clear liquid. The object is no longer visible. What could explain this effect?

14 Deduce At sunset, the Sun is at position A, but the observer sees the Sun at position B. Why does this happen?

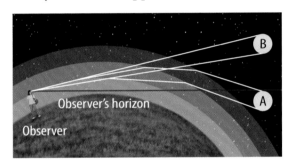

15 Construct a diagram showing how light forms an image in a concave mirror when the object is farther from the mirror than the focal point is. Explain the type of image that forms.

16 Predict what color a white shirt would appear to be if the light reflected from the shirt passes through a red filter and then a green filter before reaching your eye.

Writing in Science

17 Write You are an eye doctor with a patient who has problems seeing things up close. The patient's distance vision is good. Describe the patient's problem in detail. Identify how you would recommend solving the patient's problem.

REVIEW THE BIG IDEA

18 Describe how the interaction of matter and light affects what you see when you look at a window, a lake, and a tree.

19 How does matter affect the way you perceive light in the picture below?

Math Skills

Review — Math Practice

Use Scientific Notation

20 Light travels through a manufactured transparent material at about 1.0×10^8 m/s. What is the index of refraction of the material?

21 The speed of light is 1.83×10^8 m/s through a material. What is the material's index of refraction?

22 Light travels through a material at a speed of 1.42×10^8 m/s. What is the index of refraction of the material?

23 The speed of light in a sodium chloride crystal is 1.95×10^8 m/s. What is the index of refraction of sodium chloride?

24 A diamond's index of refraction is about 2.4. What is the speed of light through a diamond? [*Hint:* The speed of light through a substance equals the speed of light through a vacuum divided by the index of refraction.]

Standardized Test Practice

Record your answers on the answer sheet provided by your teacher or on a sheet of paper.

Multiple Choice

1. Which describes one way concave mirrors and convex mirrors differ?
   A Concave mirrors are curved, while convex mirrors are flat.
   B Concave mirrors always produce a virtual image, while convex mirrors can produce a real image.
   C Convex mirrors are curved, while concave mirrors are flat.
   D Convex mirrors always produce a virtual image, while concave mirrors can produce a real image.

2. What happens to most of the light that strikes a flat transparent object?
   A It is absorbed by the object.
   B It is magnified by the object.
   C It is reflected by the object.
   D It is transmitted by the object.

Use the figure below to answer question 3.

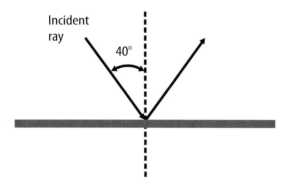

3. The figure shows how light is affected by a plane mirror. What angle does the reflected ray form with the normal?
   A 20°
   B 40°
   C 80°
   D 90°

Use the figure below to answer question 4.

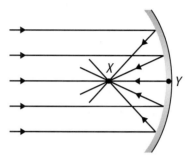

4. The figure shows how light rays are affected by a concave mirror. What is between points X and Y?
   A the angle of incidence
   B the focal length
   C the focal point
   D a virtual image

5. Which correctly describes an optical fiber?
   A The core is opaque and luminous.
   B The core is opaque and illuminates other objects.
   C The core is transparent and luminous.
   D The core is transparent and transmits light.

6. Which describes how a refracting telescope and a reflecting telescope differ?
   A A refracting telescope gathers light using a concave lens, while a reflecting telescope uses a convex lens.
   B A refracting telescope gathers light using a concave mirror, while a reflecting telescope uses a convex lens.
   C A refracting telescope gathers light using a convex lens, while a reflecting telescope uses a concave mirror.
   D A refracting telescope gathers light using a concave mirror, while a reflecting telescope uses a convex mirror.

Standardized Test Practice

7 Which describes how concave lenses and convex lenses differ?

  A Concave lenses are flat on both sides. Convex lenses are curved on both sides.

  B Concave lenses are curved on both sides. Convex lenses are flat on both sides.

  C Concave lenses are thicker in the middle. Convex lenses are thinner in the middle.

  D Concave lenses are thinner in the middle. Convex lenses are thicker in the middle.

Use the figure below to answer question 8.

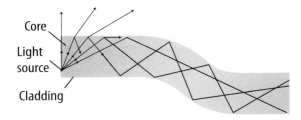

8 What determines whether light that enters the fiber will reflect back into the core?

  A the angle at which the light strikes

  B the intensity of the light

  C the thickness of the fiber

  D the wavelength of the light

9 What is unique about light from a laser?

  A It is more spread out.

  B It is more coherent.

  C It has more wavelengths.

  D Its intensity decreases faster.

Constructed Response

10 Compare and contrast the interaction of light with the rods and cones in the human eye.

11 An object viewed in white light appears to have blue and white stripes. Describe how those same stripes would appear if red light were shining on the object.

Use the figure below to answer questions 12 and 13.

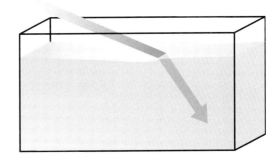

12 The figure shows light shining through air into water. What causes the light to change direction as it enters the water?

13 What would happen to the light if the tank were filled with a liquid that had the same index of refraction as air?

NEED EXTRA HELP?													
If You Missed Question...	1	2	3	4	5	6	7	8	9	10	11	12	13
Go to Lesson...	2	1	3	3	4	3	3	4	4	4	3	3	3

Chapter 19

Electricity

THE BIG IDEA How do electric circuits and devices transform energy?

Inquiry What did he spark?

You might not know the name Nikola Tesla, despite his great achievements in electrical science. In the past, many people believed electric energy was too dangerous to use in their homes. Tesla made this photograph to demonstrate that his inventions in electricity were safe for everyday use. His work set the stage for the power and lighting systems now used around the world.

- What might happen if any of the electric streaks touch the man?
- What different ways is electricity used in the room?
- How is energy transformed in this scene?

Get Ready to Read

What do you think?
Before you read, decide if you agree or disagree with each of these statements. As you read this chapter, see if you change your mind about any of the statements.

1. Protons and electrons have opposite electric charges.
2. Objects must be touching to exert a force on each other.
3. When electric current flows in a wire, the number of electrons in the wire increases.
4. Electrons flow more easily in metals than in other materials.
5. In any electric circuit, current stops flowing in all parts of the circuit if a connecting wire is removed or cut.
6. The light energy given off by a flashlight comes from the flashlight's batteries.

ConnectED Your one-stop online resource
connectED.mcgraw-hill.com

- Video
- Audio
- Review
- Inquiry
- WebQuest
- Assessment
- Concepts in Motion
- Multilingual eGlossary

Lesson 1

Reading Guide

Key Concepts
ESSENTIAL QUESTIONS

- How do electrically charged objects interact?
- How can objects become electrically charged?
- What is an electric discharge?

Vocabulary

static charge p. 680
electric insulator p. 682
electric conductor p. 682
polarized p. 683
electric discharge p. 685
grounding p. 685

 Multilingual eGlossary

 Video

- BrainPOP®
- Science Video
- What's Science Got to do With It?

Electric Charge and Electric Forces

Inquiry Where is the charge?

You are surrounded by electric charges at all times. Without them, you could not exist. However, some electric charges, such as those in lightning, can be dangerous. In what ways have electric currents been used to make your life better?

Inquiry Launch Lab

10 minutes

How can you bend water?

You open the clothes dryer, grab a warm sweater, and pull it on over your head. Then, you look in the mirror and see a sock clinging to your sleeve. A force causes the sock to cling to your sweater. What else could that force do?

1. Read and complete a lab safety form.
2. Inflate a **balloon**, and tie the end. With a **permanent marker**, draw an *X* on one side of the balloon.
3. Your partner holds a **funnel** over a **large bowl** and pours a cup of water through the funnel. As the water gently flows, bring the balloon as close to the stream of water as you can without getting it wet. Record your observations in your Science Journal.
4. Next, rub the X side of the balloon on your **sweater**, and then hold the balloon next to the spot where you rubbed. Observe the interaction between your sweater and the balloon. Record your observations.
5. Rub the X side of the balloon against your sweater again. Immediately repeat step 3.

Think About This

1. **Infer** Why did you rub the balloon on your sweater? Predict what might have happened if you simply touched the balloon to your sweater instead of rubbing it.

2. **Key Concept** Why do you think the balloon interacted the way it did with your sweater and with the stream of water?

Electric Charges

Imagine a hot summer afternoon. Dark clouds fill the sky. Suddenly, a bolt of lightning streaks across the sky. Seconds later, you hear thunder in the distance. The lightning released a tremendous amount of energy. Some of the energy was released as the light that flashed through the sky. Some of the energy was released as the sound you heard as thunder. And some of the energy was released as thermal energy that heated the air. Where did the lightning's energy come from? The answer has to do with electric charge.

Charged Particles

Recall that all matter is made of particles called atoms. Also recall that atoms are made of even smaller particles—protons, neutrons, and electrons. Protons and neutrons make up the nucleus of an atom, as shown in **Figure 1**. Electrons move around the nucleus. Protons and electrons have the property of electric charge, neutrons do not. As you read further, you will learn how charged particles interact to affect your everyday life.

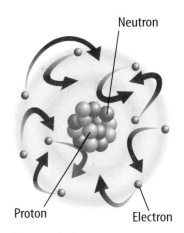

Figure 1 Protons, neutrons, and electrons make up an atom.

Reading Check Which particles found in atoms have the property of electric charge?

Lesson 1 • **679**
EXPLORE

Positive Charge and Negative Charge

There are two types of electric charge—positive charge and negative charge. Protons in the nuclei of atoms have positive charge. Electrons moving around a nucleus have negative charge. The amount of positive charge of one proton is equal to the amount of negative charge of one electron.

Recall from previous chapters that oppositely charged particles attract each other. Similarly charged particles repel each other. Therefore, a positively charged proton and a negatively charged electron attract each other. Two protons, or two electrons, repel, or push away from each other.

An atom is electrically neutral—it has equal amounts of positive charge and negative charge. This is because an atom has equal numbers of protons and electrons. Electrons can move from one atom to another. When an atom gains one or more electrons, it becomes negatively charged. If an atom loses one or more electrons, it becomes positively charged. Electrically charged atoms are called ions. It is important to understand that any object can become electrically charged.

Neutral Objects

Similar to electrically neutral atoms, larger electrically neutral objects have equal amounts of positive and negative charge. Electrically neutral objects do not attract or repel each other.

Charged Objects

Just as atoms sometimes gain or lose electrons, larger objects can gain or lose electrons, too. Some materials hold electrons more loosely than other materials. As a result, electrons often move from one object to another. When this happens, the positive charge and negative charge on the objects are unbalanced. *An unbalanced negative or positive electric charge on an object is sometimes referred to as a* **static charge.**

Like an atom, an object that gains electrons has more negative charge than positive charge. It is said to be negatively charged. Likewise, an object that loses electrons has more positive charge than negative charge. It is said to be positively charged.

Electric Forces

The region surrounding a charged object is called an electric field. An electric field applies a force, called an electric force, to other charged objects, even if the objects are not touching.

The electric force applied by an object's electric field will either attract or repel other charged objects. **Figure 2** shows that objects with opposite electric charges attract each other. On the other hand, objects with similar electric charges repel each other.

 Key Concept Check How do electrically charged objects interact?

Figure 2 Charged objects can attract or repel each other. The arrows show the direction of the objects' motion.

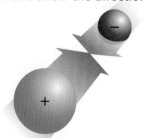

Positively and negatively charged objects attract each other.

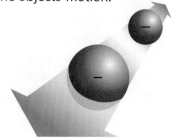

Two negatively charged objects repel each other.

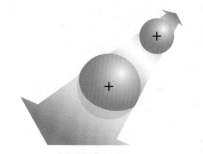

Two positively charged objects repel each other.

Electric Force and Amount of Charge

The strength of the electric force between two charged objects depends on two variables—the total amount of charge on both objects and the distance between the objects. For example, when you brush your hair, electrons move from the brush to your hair. This causes the brush and your hair to each have a static charge. The brush is positively charged. Your hair is negatively charged. Because the brush and your hair have opposite electric charges, they attract each other.

The left portion of **Figure 3** illustrates how the amount of charge affects electric force. If you brush your hair once or twice, some electrons transfer from the brush to your hair. The force of attraction is not very strong. However, if you continue brushing your hair, more electrons transfer from the brush to your hair. This increases the strength of the electric force of attraction between the brush and your hair.

Electric Force and Distance

Distance also affects the strength of the electric force between electric charges. As described above, brushing your hair produces an unbalanced positive charge on the brush and an unbalanced negative charge on your hair. Electric fields surround these electric charges. The fields are more intense near the charges. A more intense electric field applies a stronger electric force to other objects. Therefore, when the brush is close to your hair, the force of attraction is very noticeable. The electric force is weaker when the charged objects are far from each other, as shown on the right side of **Figure 3**.

FOLDABLES
Make a vertical three-tab book, and label it as shown. Use it to organize your notes about electrically charged objects.

Figure 3 The strength of the electric force between two charged objects depends on the distance between the objects and the total amount of electric charge of the objects.

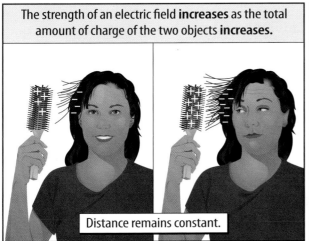

The strength of an electric field **increases** as the total amount of charge of the two objects **increases**.

Distance remains constant.

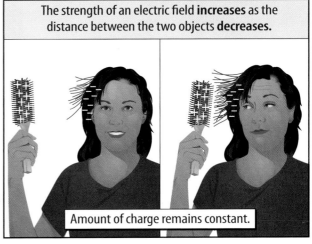

The strength of an electric field **increases** as the distance between the two objects **decreases**.

Amount of charge remains constant.

MiniLab — 10 minutes

How can a balloon push or pull?

When a balloon touches a wool cloth, electrons transfer from the wool onto the balloon.

1. Read and complete a lab safety form.
2. Inflate and tie off two **balloons.** Attach a 10 cm length of **string** to each balloon.
3. Hold the balloons by the strings and slowly bring them together. Record your observations in your Science Journal.
4. Hold one of the balloons 1–3 cm over some **packing peanuts.** Record your observations.
5. Now rub each balloon against a piece of **wool cloth.** Repeat step 3.
6. Rub one of the balloons against the wool again. Repeat step 4.

Analyze and Conclude

1. **Construct** a diagram showing the arrangement of the charges as you are rubbing the balloon against the wool cloth, when the two balloons are interacting, and when the balloon is held near the packing peanuts.
2. **Key Concept** Summarize how objects with electric charge interact.

Transferring Electrons

You read that a hair brush and strands of hair become electrically charged when electrons transfer from the brush to your hair. How do electrons move from one object to another?

Insulators and Conductors

To understand how electrons move from one object to another, you need to know about two basic types of materials—electric insulators and electric conductors. *A material in which electrons cannot easily move is an* **electric insulator.** Glass, rubber, wood, and even air are good electric insulators. *A material in which electrons can easily move is an* **electric conductor.** Most metals, such as copper and aluminum, are good electric conductors.

Electric insulators and electric conductors are all around you. Electric conductors and electric insulators are used in electrical power cords around your house. In **Figure 4,** copper wire is the conductor of electrons in an extension cord. The copper allows electrons to easily move through the wire. Plastic and rubber are used as protective electric insulators around the metal wire. Electrons cannot easily move in the plastic and rubber.

Electrons transfer between objects by contact, induction, or conduction. As you will read, electric insulators and conductors play an important role in these processes.

Reading Check How do electric insulators and electric conductors differ?

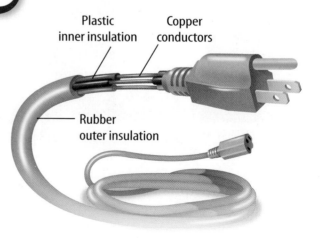

Figure 4 Using conductors and insulators is one way to safely control the flow of electric charges.

Review   Personal Tutor

Transferring Charge by Contact

Recall that some materials hold their electrons more loosely than other materials. When objects made of different materials touch, electrons tend to collect on the object that holds electrons more tightly. This is called transferring charge by contact.

The wool sweater in **Figure 5** holds electrons more loosely than the rubber balloon. When the balloon comes in contact with the sweater, electrons from the surface of the sweater transfer to the surface of the balloon, creating a static charge on both objects. Because the balloon gained electrons, it has an unbalanced negative charge. Because the sweater lost electrons, it has an unbalanced positive charge. Both insulators and conductors can be charged by contact.

Transferring Charge by Induction

Transferring charge by induction is a process by which one object causes two other objects that are conductors to become charged without touching them. **Figure 6** illustrates how this works.

Part 1 of **Figure 6** shows a negatively charged balloon repelling electrons in a metal soda can. Because the can is aluminum and, therefore, a conductor, electrons in the can easily move toward the far end of the can. The can is not charged because it has not gained or lost any electrons. Instead, the can is polarized. *When electrons concentrate at one end of an object, the object is* **polarized.**

In part 2, when a charged balloon is brought near two cans that are touching, the balloon polarizes the cans as if they are one object. Electrons in both cans move toward the far end of the can on the right. Then, the two cans separate. As shown in part 3, the cans that were originally polarized as a group are now individually charged. The can on the right has an unbalanced negative electric charge, and the can on the left has an unbalanced positive electric charge.

▲ **Figure 5** More loosely held electrons on wool easily transfer onto the surface of a rubber balloon.

Figure 6 Objects that are electric conductors can be charged by induction. ▼

❶

❷

❸
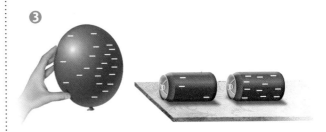

Visual Check Why do negative charges in the aluminum cans tend to move away from the balloon?

Transferring Charge by Conduction

Another way that electrons transfer between two conductors is called transferring charge by conduction. As shown in **Figure 7,** when conducting objects with unequal charges touch, electrons flow from the object with a greater concentration of negative charge to the object with a lower concentration of negative charge. This is similar to water flowing from a container with a higher water level to a container with a lower water level. The flow of electrons continues until the concentration on charge of both objects is equal.

 Key Concept Check What are three ways by which objects can become electrically charged?

Figure 7 Water flows from the container with more water to the container with less water until the levels are equal. Similarly, negative charges flow between objects until the concentration of charge on both objects is equal.

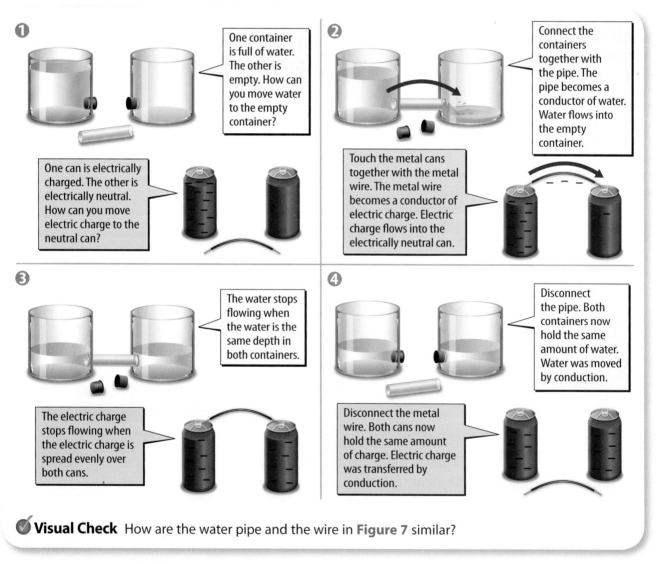

Visual Check How are the water pipe and the wire in **Figure 7** similar?

Electric Discharge

When you brush your hair, electrons transfer from the brush to your hair. What happens to these excess electrons in your hair? They transfer to objects that come in contact with your hair, such as a hat or your pillow. Electrons even transfer to the air. Gradually, your hair loses its charge. *The process of an unbalanced electric charge becoming balanced is an* **electric discharge.**

Lightning Rods and Grounding

An electric discharge can occur slowly, such as when your hair loses its negative charge and is no longer attracted to a brush. Or, an electric discharge can occur as quickly as a flash. For example, lightning is a powerful electric discharge that occurs in an instant.

A lightning strike can severely damage a building or injure people. Lighting rods help protect people against these dangers. **Figure 8** shows a metal lightning rod attached to the roof of a building. A thick wire connects the lightning rod to the ground. The wire provides a path for the electrons released in a lightning strike to travel into the ground. *Providing a path for electric charges to flow safely into the ground is called* **grounding.** **Table 1** provides tips on how to protect yourself from a lightning strike.

What causes lightning?

Scientists do not fully understand what causes lightning. However, many scientists think that lightning is related to the electric charges that separate within storm clouds. The large amounts of ice, hail, and partially frozen water droplets that thunderstorms create seem to play a role, too.

Forecasting exactly where and when lightning will strike might never be possible. **Figure 9** on the next page summarizes what is known about how lightning forms.

 Key Concept Check What is an electric discharge?

Figure 8 Lightning rods and proper grounding help protect tall buildings from the damaging effects of lightning.

Visual Check What are the two components of a building's lightning protection system?

WORD ORIGIN
electric
from Greek *ēlektron*, means "beaming sun"

Table 1 Lightning Safety Tips

If the time between a lightning flash and thunder is less than 30 seconds, the storm is dangerously close. Protect yourself in the following ways:
- Seek shelter in an enclosed building or a car with a metal top. Never stand under a tree.
- Do not touch metal surfaces.
- Get away from water if swimming, boating, or bathing.
- Wait 30 minutes after the last flash of lightning before leaving the shelter, even if the Sun comes out.

Figure 9 The conditions needed for lightning to form are well known. However, scientists still do not fully understand how lightning is produced. Lightning is one of the most unpredictable forces of nature.

1 Within a storm cloud, warm air rises past falling cold air. The cold air is filled with hail, ice, and partially frozen water droplets that pick up electrons from the rising air. This causes the bottom of the cloud to become negatively charged.

2 The negatively charged cloud polarizes Earth's surface by repelling negative charges in the ground. Thus, the surface of the ground becomes positively charged.

3 When the bottom of the cloud accumulates enough negative charge, the attraction of the positively charged ground causes electrons in the cloud to begin moving toward the ground.

4 As electrons approach the ground, positive charges quickly flow upward, making an electric connection between the cloud and the ground. You see this electric discharge as lightning.

5 Accumulation of charge does not only occur between clouds and the ground. Charges also separate within or between storm clouds, causing cloud-to-cloud lightning.

Lesson 1 Review

Assessment Online Quiz

Visual Summary

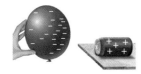
Any object can be positively charged, negatively charged, or neutral.

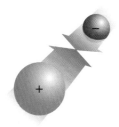
Charged objects exert forces on each other.

Lightning is one form of electric discharge.

FOLDABLES
Use your lesson Foldable to review the lesson. Save your Foldable for the project at the end of the chapter.

What do you think NOW?

You first read the statements below at the beginning of the chapter.

1. Protons and electrons have opposite electric charges.
2. Objects must be touching to exert a force on each other.

Did you change your mind about whether you agree or disagree with the statements? Rewrite any false statements to make them true.

Use Vocabulary

1. Providing a path for electric charges to flow safely into the ground is called _____.

2. **Distinguish** between an electric conductor and an electric insulator.

3. **Use the term** *static charge* in a sentence.

Understand Key Concepts

4. Particles that attract each other are
   A. two electrons.
   B. two protons.
   C. one proton and one electron.
   D. one proton and one neutron.

5. **Summarize** How does the electric force between charged objects change as the objects move away from each other?

6. **Describe** Imagine you have to design a safe, useful device that uses an electric discharge to make your life better. How would you describe it to others?

Interpret Graphics

7. **Organize** Copy the graphic organizer below. Fill in the type of charge on protons, neutrons, and electrons.

Particle	Type of Electric Charge
Proton	
Neutron	
Electron	

Critical Thinking

8. **Analyze** How can objects become positively charged?

9. **Hypothesize** A polyester shirt clings to your skin as you put it on. You then learn that skin easily releases electrons. Write a description of what happened between the shirt and your skin.

HOW IT WORKS
Van de Graaff Generator

How can a machine move electrons?

Have you ever seen a Van de Graaff generator? Originally, scientists studied the nuclei of atoms with these machines. Now, they are often seen in science museums, where they are used to demonstrate the effects electric charge.

How does this device generate electric charge? Recall that some materials hold electrons more loosely than other materials. In a Van de Graaff generator, the metal dome holds electrons more loosely than the belt. As the belt travels over the top roller past the upper comb, electrons move from the metal dome through the upper comb and onto the belt. This leaves the dome positively charged.

The excess electrons on the belt move from the belt onto the lower comb as the belt moves over the roller at the bottom of the generator. The electrons then travel through a wire out of the machine and into the ground. This process of electrons moving from the dome to the ground continues as long as the generator is running.

Soon, the dome loses so many electrons that it acquires a very large positive charge. The positive charge on the dome becomes so great that electrons create an electric spark as they jump back onto the dome from any object that will release them. That object could be you if you stand close enough.

▼ The Van de Graaff generator causes the girl's body, including her hair, to become electrically charged. Because all the strands of her hair acquire the same charge and repel each other, her hair stands on end.

Upper comb
Metal dome
Belt
Electric motor
Lower comb
Ground

It's Your Turn

REPORT Find out how to build a simple Van de Graaff generator using everyday materials. Share your findings with your classmates.

Lesson 2

Reading Guide

Key Concepts
ESSENTIAL QUESTIONS

- What is the relationship between electric charge and electric current?
- What are voltage, current, and resistance? How do they affect each other?

Vocabulary
electric current p. 690
electric circuit p. 690
electric resistance p. 692
voltage p. 693
Ohm's law p. 694

 Multilingual eGlossary

 Video

- BrainPOP®
- Science Video

Electric Current and Simple Circuits

Inquiry) How far can it go?

Across the country, over 300,000 km of wires carry electric energy from about 500 power companies to homes and businesses. A flashlight operates with only a few centimeters of electric conductors. How are these vastly different systems similar?

Inquiry Launch Lab

20 minutes

What is the path to turn on a light?
How can you make a lightbulb light up?

1. Read and complete a lab safety form.
2. Examine a **D-cell battery.** Notice the differences between the two ends. Record your observations in your Science Journal.
3. Using a **paper clip, small lightbulbs,** and the battery, design a way to light the bulb. Draw a diagram and test your plan.
   ⚠ *The paper clip can get hot!*
4. Find two other ways to light the bulb using the paper clip and the battery. Draw and test your plans.
5. Record three configurations that do not light the bulb.

Think About This

1. Why do some arrangements of the materials light the bulb, but other arrangements do not?
2. 🔑 **Key Concept** How do you think all the pieces interact to light the bulb?

SCIENCE USE V. COMMON USE

current
Science Use the flow of electric charge

Common Use occurring at the present time

WORD ORIGIN

circuit
from Latin *circuire*, means "to go around"

Electric Current and Electric Circuits

In Lesson 1, you read about ways to transfer, or move electric charges. The transferring of electric charges allows you to power the electrical devices you use every day. For example, conduction of electric charges occurs every time you turn on a flashlight or a television. Induction is used for wireless charging of devices, such as electrical toothbrushes. Transferring charge by contact also occurs every time lightning strikes. How are a lightning flash and a TV similar? They both transform the energy of moving electrons to light, sound, and thermal energy. *The movement of electrically charged particles is an* **electric current.**

A Simple Electric Circuit

The movement of electrons in a lightning strike lasts only a fraction of a second. In a TV, electrons continue moving as long as the TV remains on. An electric current flows in a closed path to and from a source of electric energy. *A closed, or complete, path in which an electric current travels is an* **electric circuit.**

All electric circuits have one thing in common—they transform electric energy to other forms of energy. For example, electric circuits in a microwave oven transform electric energy to the thermal energy that quickly cooks your food. Circuits in your television transform electric energy to the light and sound that entertains and informs you. How do electric energy transformations improve your life?

▲ **Figure 10** An electric current flows in a circuit if the circuit is complete, or closed. Current will not flow if the circuit is broken, or open.

How Electric Charges Flow in a Circuit

Figure 10 shows a battery, wires, and a lightbulb connected in a circuit. The lightbulb glows as electrons flow from of one end of the battery, through the wires and lightbulb, and back into the other end of the battery. If the circuit is broken, or open, the electrons stop flowing, and the bulb stops glowing.

A current of electrons in a wire is somewhat like water being pumped through a hose. The amount of water flowing into one end of a hose is the same as the amount of water flowing out the other end. As shown in **Figure 11,** the number of electrons flowing into a wire from a power source equals the number of electrons flowing out of the wire back into the source.

▲ **Figure 11** When current flows in a wire, the number of electrons in the wire does not change.

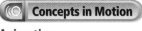

The Unit For Electric Current

Electric current is approximately measured as the number of electrons that flow past a point every second. However, electrons are so tiny and there are so many in a circuit that you could never count them one at a time. Therefore, just like you count eggs by the dozen, scientists count electrons by a quantity called the coulomb (KEW lahm). A coulomb is a very large number—about 6×10^{18}, or 6,000,000,000,000,000,000! That is 6 quintillion.

The SI unit for electric current is the ampere (AM pihr), commonly called an amp. Its symbol is A. One ampere of current equals about one coulomb of electrons flowing past a point in a circuit every second. The electric current through a 120-W lightbulb is about 1 A. A typical hair dryer uses a current of about 10 A, or 60,000,000,000,000,000,000 electrons per second.

 Key Concept Check What is the relationship between electric charge and electric current?

Inquiry MiniLab 20 minutes

When is one more than two?

All parts of a circuit—such as bulbs and wires—add resistance to the circuit.

1. Read and complete a lab safety form.
2. Using **alligator clip wires,** connect a **small bulb in a base** to a **hand-cranked generator.** Draw your circuit in your Science Journal.
3. Crank the generator at different rates. Observe what happens to the brightness of the bulb as the crank turns. Turn the crank at a rate that produces a medium brightness. Record that rate.
4. Using additional wires as needed, add a **second bulb** to your circuit. Reconnect the generator, and turn the crank at the same rate as before. Draw the circuit and record your observations.
5. Repeat the previous step with additional bulbs and wires.

Analyze and Conclude

1. **Describe** how you changed the resistance in the circuit.
2. **Key Concept** Explain how current and resistance are related in this circuit.

Table 2 Electric Resistance of Different Materials	
Material (20cm x 1mm)	Resistance (ohms)
Copper	0.004
Gold	0.006
Iron	0.025
Carbon	8.9
Rubber	2,500,000,000,000,000,000

What is electric resistance?

See **Figure 12** on the next page. Suppose you replaced one of the green wires in the circuit on the left side with a piece of string. The lightbulb would not glow because there would be no current in the circuit. Why do electrons flow easily in metal wire but not in string? The answer is that wire has much less electric resistance than string. **Electric resistance** *is a measure of how difficult it is for an electric current to flow in a material.*

The unit of electric resistance is the ohm (OHM), which is symbolized by the Greek letter Ω (omega). An electric resistance of 20 ohms is written 20 Ω. **Table 2** lists the electric resistances of different materials.

Electric Resistance of Conductors and Insulators

You read that electric conductors are materials, such as copper and aluminum, in which electrons easily move. A good conductor has low electric resistance. Usually electric wires are made of copper because copper is one of the best conductors.

Recall that electrons cannot easily move through insulators such as plastic, wood, or string. A good electric insulator has high electric resistance. Atoms of an insulator hold electrons tightly. This prevents electric charges from easily moving through the material. Therefore, replacing a wire in a circuit with string prevents electrons from flowing in a circuit.

Resistance—Length and Thickness

A material's electric resistance also depends on the material's length and thickness. A thick copper wire has less electric resistance than a thin copper wire of the same length. Because the thick wire has less resistance, it will conduct better. Increasing the length of a conductor also increases its electric resistance.

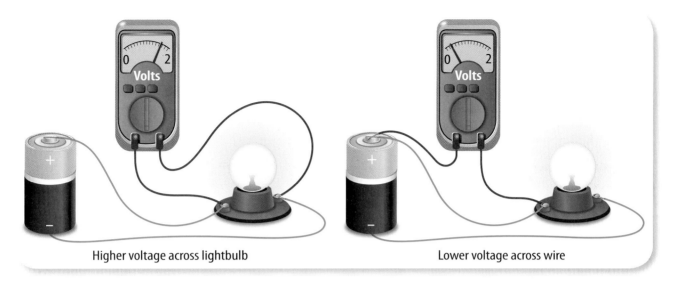

Higher voltage across lightbulb — Lower voltage across wire

What is voltage?

You probably heard the term *volt*. You use 1.5-V batteries in a flashlight. You plug a hair dryer into a 120-V outlet. But, what does this mean?

Battery Voltage

In **Figure 12**, a battery creates an electric current in a closed circuit. Energy stored in the battery moves electrons in the circuit. As the electrons move through the circuit, the amount of energy transformed by the circuit depends on the battery's voltage. **Voltage** *is the amount of energy the source uses to move one coulomb of electrons through the circuit.* A circuit with a high voltage source transforms more electric energy to other energy forms than a circuit with a low voltage source. For example, a lightbulb connected to a 9-V battery produces about six times more light and thermal energy than the same lightbulb connected to a 1.5-V battery.

Voltage in Different Parts of a Circuit

Electric energy transforms to other forms of energy in all parts of a circuit. For example, the lightbulbs in **Figure 12** transform electric energy to light and thermal energy. Even the wires and batteries produce a small amount of thermal energy. In other words, different amounts of energy transform in different parts of a circuit. The voltage measured across a portion of a circuit indicates how much energy transforms in that portion of the circuit. For example, the voltage is greater across the lightbulb in **Figure 12** than across the wire. Therefore, the lightbulb transforms more energy than the wire.

Reading Check What happens to the energy flowing in an electric circuit?

Figure 12 Voltage can be different in different parts of a circuit. The voltmeter shows where most of the battery's energy is used.

Visual Check Why is the voltage reading across the lightbulb higher than across the segment of wire?

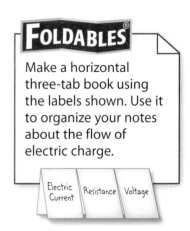

Make a horizontal three-tab book using the labels shown. Use it to organize your notes about the flow of electric charge.

Ohm's Law

When designing electrical devices, engineers choose materials based on their electric resistance. For example, the heating coils in a toaster must be made of a metal with very high electric resistance. As explained on the next page, this allows the coils to transform most of the circuit's energy to thermal energy. But, how much resistance should a conductor have? The answer is found with Ohm's law.

Using Ohm's Law

Named after German physicist Georg Ohm, **Ohm's law** *is a mathematical equation that describes the relationships among voltage, current, and resistance.* The law states that as the voltage of a circuit's electric energy source increases, the current in the circuit increases, too. Also, as the resistance of a circuit increases, the current decreases.

Ohm's Law can be written as the following equation:

> **Ohm's Law Equation**
>
> voltage (V) = current (I) × resistance (R)
>
> $V = IR$

V is the symbol for voltage, measured in volts (V). I is the symbol for electric current, which is measured in amperes (A). And, R is the symbol for electric resistance, measured in ohms (Ω). If you know the value of two of the variables in the equation, you can determine the third, as described in **Figure 13**. You can measure any of these variables with a multimeter, as shown in **Figure 14**.

 Key Concept Check How do voltage, current, and resistance affect each other?

Figure 13 Ohm's law can be used to calculate unknown quantities in a circuit.

Calculate the voltage across the lightbulb.

$R = 50Ω$
$I = 0.1A$
$V = ?$

To find the voltage, start with Ohm's law:

$V = IR$

Substitute the known values into the equation:

$V = 0.1A \times 50Ω$
$V = 5V$

Calculate the resistance of the lightbulb.

$R = ?Ω$
$I = 0.3A$
$V = 6.0V$

To find the resistance, start with this form of Ohm's law:

$R = \dfrac{V}{I}$

Substitute the known values into the equation:

$R = \dfrac{6V}{0.3A}$
$R = 20Ω$

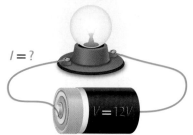

Calculate the current flowing through the lightbulb.

$R = 50Ω$
$I = ?$
$V = 12V$

To find the current, start with this form of Ohm's law:

$I = \dfrac{V}{R}$

Substitute the known values into the equation:

$I = \dfrac{12V}{50Ω}$
$I = 0.2A$

Math Skills — Use a Simple Equation

Solve for Voltage The current through a lightbulb is **0.5 A**. The resistance of the lightbulb is **220 Ω**. What is the voltage across the lightbulb?

❶ This is what you know: current: $I = 0.5$ A
  resistance: $R = 220$ Ω

❷ This is what you need to find: voltage V

❸ Use this formula: $V = IR$

❹ Substitute: $V = (0.5\ \text{A}) \times (220\ \Omega) = 110\ \text{V}$
the values for I and R into the formula and multiply

Answer: The voltage is **110 V**.

Practice
What is the voltage across the ends of a wire coil in a circuit if the current in the wire is 0.1 A and the resistance is 30 Ω?

- Review
- Math Practice
- Personal Tutor

Voltage, Resistance, and Energy Transformation

Figure 14 shows two lightbulbs and a battery connected in a circuit. Both bulbs are connected one-after-another in a single loop. The currents through both bulbs are equal. However, the two lightbulbs are not identical. One has greater electric resistance than the other.

You determine which lightbulb has more electric resistance with a voltmeter and an understanding of Ohm's law. Ohm's law states that, with equal current, the voltage is greater across the device with the greater resistance. Figure 14 also shows that the lightbulb with the greater electric resistance has the greater voltage across it. The higher-resistance lightbulb on the left transforms more electric energy to light.

REVIEW VOCABULARY
energy transformation
the conversion of one form of energy to another

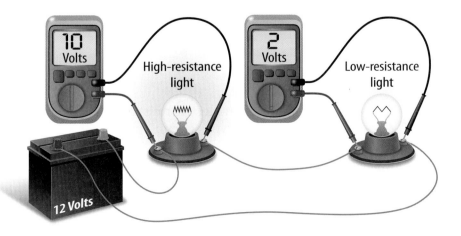

Figure 14 When two devices are in the same circuit, the voltage is greater across the device with higher resistance. The device with higher resistance transforms more electric energy to other forms of energy.

Visual Check Which lightbulb transforms more electric energy?

Lesson 2 Review

Visual Summary

Over 300,000 km of wires carry electric current to homes and businesses.

Electric circuits transform electric energy to other forms of energy.

Ohm's law shows the relationship among voltage, current, and resistance.

FOLDABLES

Use your lesson Foldable to review the lesson. Save your Foldable for the project at the end of the chapter.

What do you think NOW?

At the beginning of this lesson, you read the statements below.

3. When electric current flows in a wire, the number of electrons in the wire increases.

4. Electrons flow more easily in metals than in other materials.

Did you change your mind about whether you agree or disagree with the statements? Rewrite any false statements to make them true.

Use Vocabulary

1. Electrons flow more easily in a material with lower _____.

2. A closed path in which electric charges can flow is a(n) _____.

3. The flow of electric charges is a(n) _____.

Understand Key Concepts

4. Ohm's law is NOT related to
  A. current.   C. resistance.
  B. mass.   D. voltage.

5. Describe in your own words the relationship between electric charge and electric current.

Interpret Graphics

6. Explain Copy the table below. Fill in the effect of changing voltage and resistance on electric current.

Variable	Effect of Increase in Variable on Current
Voltage	
Resistance	

Critical Thinking

7. Evaluate How does the total electric charge of a wire in a circuit change when current stops flowing in the circuit?

Math Skills

Math Practice

8. The current in a hair dryer is 10 A. The hair dryer is plugged into a 110-V outlet. How much electric resistance is there in the hair dryer's circuit?

Inquiry Skill Practice: Identify and Manipulate Variables — 40 minutes

What effect does voltage have on a circuit?

Materials

2 batteries in bases

2, 3-V bulbs in bases

alligator clip wires

Safety

In 1827, German scientist Georg Ohm published his work about the relationship among voltage, current, and resistance. At that time many people and other scientists disagreed with him. They did not think that experiments were necessary to understand the world. We now know that the best way to develop an understanding of nature is through carefully developed and controlled experiments, such as those Ohm performed, in which the researcher **identifies and manipulates the variables.**

Learn It

In an experiment, it is important to keep all factors the same except for the factor you are testing. The factor you manipulate, or change, is called the **independent variable.** The factor that changes as a result is the **dependent variable.**

Try It

1. Read and complete a lab safety form.

2. Notice how the two ends of the battery differ. Record your observations in your Science Journal.

3. Copy the table below in your Science Journal. Record your predictions for each circuit.

4. Construct and observe the circuits described in the table.

5. Compare the predicted brightness of the bulb in each circuit to the observed brightness. Record your findings.

Apply It

6. **Determine** the dependent and the independent variables.

7. **Key Concept** Describe how voltage and current relate in each circuit.

Circuit Description	Sketch of Circuit	Predicted Bulb Brightness	Observed Brightness
One battery and one bulb	(sketch)		
Two batteries and one bulb version #1			
Two batteries and one bulb version #2			

Lesson 2 EXTEND

Lesson 3

Describing Circuits

Reading Guide

Key Concepts
ESSENTIAL QUESTIONS

- What are the basic parts of an electric circuits?
- How do the two types of electric circuits differ?

Vocabulary
series circuit p. 701
parallel circuit p. 702

Multilingual eGlossary

Video   Science Video

Inquiry Can wires fly?

Modern jets may contain over 160 km of electric circuits. However, aircraft wiring does not last for the life of an airplane. Wiring systems in airplanes can wear out in 10 to 15 years. Why is it important for airplane mechanics to understand electric circuits?

Inquiry Launch Lab

20 minutes

How would you wire a house?

If you turn off the kitchen light, why does the refrigerator keep running?

1. Read and complete a lab safety form.

2. Using **alligator clip wires,** light a **small bulb in a base** with a **D-cell battery in a base.** Draw the setup in your Science Journal.

3. Disconnect the wires between the battery and the bulb. Add a **second bulb** between the battery and the first bulb. Remove one of the bulbs from its base. Record your observations.

4. Take apart the setup, and reassemble the setup in step 2. Now, add a second bulb by connecting one terminal of the second bulb to one of the terminals of the lit bulb. Connect the other terminal of the second bulb to the remaining terminal of the lit bulb. Both bulbs should be lit. Remove one of the bulbs from its base. Record your observations.

Think About This

1. Which of your assembled circuits would be the best to use when wiring a house?

2. **Key Concept** Which setup is the best way to connect lights in your home? Explain.

Parts of an Electric Circuit

Do you study by an electric lamp? Is there a computer on your desk? These **devices** contain electric circuits.

Three common parts of most electric circuits are a source of electric energy, electrical devices that transform the electric energy, and conductors, such as wires, that connect the other components.

Electric Energy to Kinetic Energy

An energy source, such as the battery in **Figure 15,** produces an electric current in a circuit. Some electrical devices are designed to transform the electric energy of the current to kinetic energy—the energy of motion. For example, the motor in an electric fan transforms electric energy to the kinetic energy of moving air particles that keep you cool.

Key Concept Check What are the three basic parts of an electric circuit?

ACADEMIC VOCABULARY
device
(noun) a piece of equipment

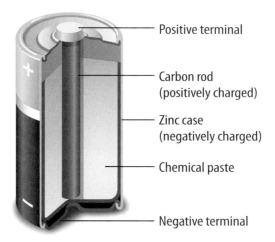

Figure 15 In a battery, chemical reactions in the moist paste cause the carbon rod to become positively charged. The zinc case becomes negatively charged.

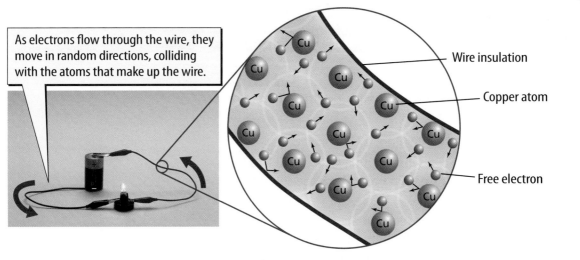

Figure 16 The battery produces a current of electrons in a closed circuit. The kinetic energy of the electrons transforms to other forms of energy.

 Animation

Batteries supply electric energy.

A battery is often used as a source of electric energy for a circuit. **Figure 15** on the previous page shows a cross section of a common battery. A battery is just a can of chemicals. As the chemicals react, electrons in the battery concentrate at the battery's negative terminal. When a closed circuit connects the terminals, electrons flow in the circuit from the battery's negative terminal to the positive terminal. If the circuit is closed and the chemicals in the battery keep reacting, an electric current continues.

Electric circuits transform energy.

Energy transformations occur in all parts of an electric circuit. For example, a battery transforms stored chemical energy to the electric energy of electrons moving as an electric current. Electrical devices in the circuit transform most of the electric energy to other useful forms of energy. For example, a lightbulb transforms electric energy to light, and stereo speakers transform electric energy to sound. In addition, all parts of a circuit, including the energy source and the connecting wires, transform some of the electric energy to wasted thermal energy.

Figure 16 shows how these energy transformations occur. For example, electrons flowing in the wire filament of a lightbulb collide with the atoms that make up the filament. The collisions transfer the electrons' energy to the atoms. The atoms immediately release the energy in other forms, such as light and thermal energy.

You read that circuits and devices release wasted thermal energy. However, many modern electrical devices, such as compact fluorescent lamps (CFLs), are designed to waste less energy. CFLs conduct an electric current through a gas and do not use wire filaments. More energy transforms to light, and less is wasted as thermal energy.

Wires connect parts of a circuit.

Recall that an electric current flows only in a closed circuit. Therefore, a circuit's energy source and device must be connected with some form of conducting material. Metal wires often are used to complete circuits. Because of their low electric resistance, wires transform only a small amount of electric energy to wasted thermal energy. This leaves more energy available for useful devices in the circuit.

Series and Parallel Circuits

Some strings of holiday lights will not light if any bulbs are missing. Other strings of lights work with or without missing bulbs. These two examples of holiday lights represent the two types of electric circuits—series circuits and parallel circuits.

Series Circuit—A Single Current Path

A **series circuit** *is an electric circuit with only one path for an electric current to follow.* Many strings of holiday lights are series circuits. In these series circuits all the lightbulbs connect end-to-end in a single conducting loop. As shown in **Figure 17,** breaking the loop at any point stops the current of electrons throughout the entire circuit. If a wire loop in a series circuit is broken, or open, all devices in the circuit will turn off.

Reading Check Why do all devices in a series circuit stop working if the circuit is broken at any point?

The amount of current in a circuit depends on the number of devices in the circuit. **Figure 17** shows that adding devices to a series circuit adds resistance to the total circuit. According to Ohm's law, when resistance increases and voltage remains the same, there is less current in the circuit. Because a string of holiday lights contains so many devices, the current through the circuit is very low. The low current produces very little thermal energy, making these lights safe to use.

WORD ORIGIN
parallel
from Greek *parallēlos*, means "beside of one another"

Figure 17 A series circuit has only one path for current. The current stops if that path is broken.

Visual Check What happens if one of the wires breaks or is cut?

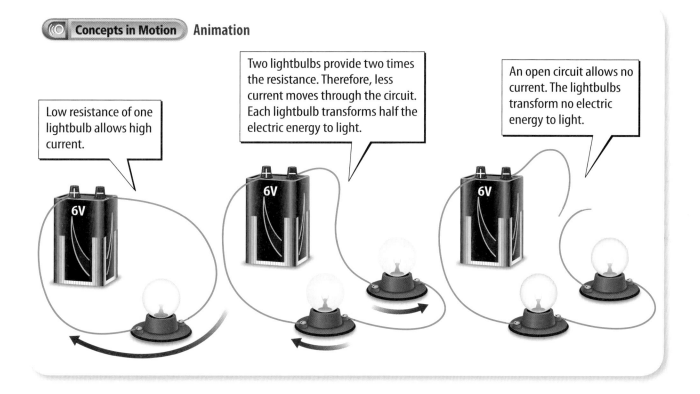

Concepts in Motion Animation

Low resistance of one lightbulb allows high current.

Two lightbulbs provide two times the resistance. Therefore, less current moves through the circuit. Each lightbulb transforms half the electric energy to light.

An open circuit allows no current. The lightbulbs transform no electric energy to light.

Figure 18 Parallel circuits have more than one path through which electric current can flow.

In a parallel circuit, each device has its own path through which current can flow.

If any of the paths of a parallel circuit is broken, current still can flow through the other devices.

However, if the circuit is open at the source of electric energy, no current can flow through any of the devices.

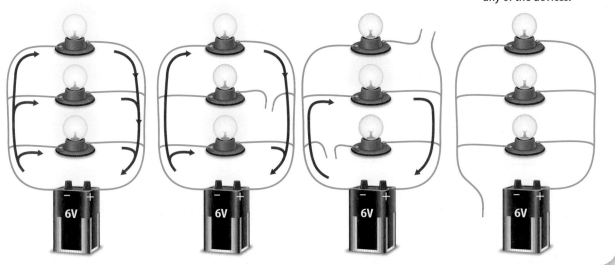

Parallel Circuit— Multiple Current Paths

If the electrical devices in your home were connected as a series circuit, you would need to turn on every electrical device in the house just to watch TV! Luckily, devices in your home are connected to an electric source as a **parallel circuit**—*an electric circuit with more than one path, or branch, for an electric current to follow.*

As shown in **Figure 18**, if you open one branch of a parallel circuit, current continues through the other branches. As a result, you can turn off the TV but keep the kitchen light and the refrigerator on.

You read that a device in a parallel circuit connects to the source with its own branch. The current in one branch has no effect on the current in other branches. However, adding branches to a parallel circuit does increase the total current through the source.

Key Concept Check How do the two types of electric circuits differ?

Electric Circuits in the Home

Most people use electrical devices every day without thinking much about them. However, where does the electric energy that we use come from?

Electric energy used in most homes and businesses is generated at large power plants. These power plants may be many kilometers from your home. A complex system of transmission wires carries the electric energy to all parts of the country.

Figure 19 shows that electric energy enters your home through a main wire. You might see this wire coming from a utility pole outside your home. Sometimes the main wire is underground. Either way, before coming into your house, the main wire travels through an electric meter. The meter measures the energy used in the electric circuits of your home.

From the meter, the wire enters your house and goes to a steel box called the main panel. At the main panel, the main wire divides into the branches of the parallel circuit that carry electric energy to all parts of your house.

702 • Chapter 19
EXPLAIN

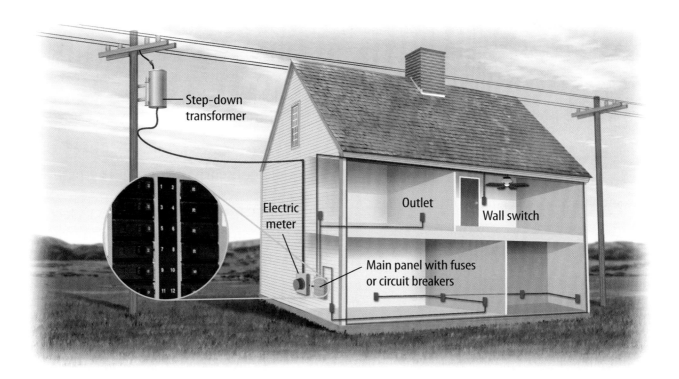

Fuses, Circuit Breakers, and GFCI Devices

Recall that adding branches and devices to a parallel circuit increases the current through the source. Too many branches or devices in a circuit can create dangerously high current in the circuit. Excess current can cause the circuit to become hot enough to cause a fire. The branches of your home's parallel circuit connect to the source with a safety mechanism, such as a fuse or a circuit breaker, to help prevent such disasters.

Fuses and circuit breakers automatically open a circuit when the current becomes dangerously high. A fuse is a piece of metal that breaks a circuit by melting from the thermal energy produced by a high current. A circuit breaker is a switch that automatically opens a circuit when the current is too great.

There is another type of automatic safety mechanism found in some circuits. Are there electric outlets in your home with two small buttons labeled *test* and *reset*? Those special outlets are ground-fault circuit interrupters (GFCI). A GFCI is used where an outlet is found near a source of water, such as a sink. A GFCI protects you from a dangerous electric shock.

For example, imagine you are in the bathroom using a hair dryer. Water accidentally splashes on you and the hair dryer. Since tap water is an electric conductor, some of the electric current from the outlet flows through you and not through the hair dryer. If a current flows through you, it could be fatal. As soon as the GFCI senses that not all of the current is flowing through the hair dryer, it opens the circuit and stops the current. It is able to react as quickly as 1/30 of a second.

Figure 19 The electric devices in a house are connected in parallel circuit. Each outlet or fixture is connected to a separate branch.

Visual Check Where are circuit breakers located in the home?

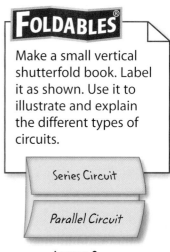

Make a small vertical shutterfold book. Label it as shown. Use it to illustrate and explain the different types of circuits.

Inquiry MiniLab 20 minutes

What else can a circuit do?

Electric circuits can do more than light a few bulbs. Look around your kitchen. Circuits transform electric energy to help you do everything from cooking dinner to freezing the leftovers to washing the dishes.

1. Copy the table below into your Science Journal.

Device	What it does	Source of electric energy	Energy transformed into

2. At each lab station, examine the device displayed. Record your observations in your table.

Analyze and Conclude

1. **Select** two devices and compare and contrast their operation.
2. **Key Concept** How do you think the circuits of the devices you examined differ? Explain your reasoning.

Electric Safety

An electric shock can be painful, and sometimes deadly. Each year, more than 500 people die by accidental electric shock in the United States.

What causes an electric shock?

An electric current follows the path of least electric resistance to the ground. That path could be through any good electric conductor, such as metal, wet wood, water, or even you! An electric shock occurs when an electric current passes through the human body. If you touch a bare electric wire or faulty appliance while you are grounded, an electric current could pass through you to the ground, resulting in a dangerous shock.

Current as small as 0.01 A can produce a painful shock. More than 0.1 A of electric current can cause death. The voltage of household electrical devices can cause dangerous amounts of current to pass through the body.

How can you be safe?

Listed below are some ways you can help protect yourself from a deadly electric shock:

- Never use electrical devices with damaged power cords.
- Stay away from water when using electrical devices plugged into an outlet.
- Avoid using extension cords and never plug more than two home appliances into an outlet at once.
- Never allow any object that you are touching to contact electric power lines, such as a kite string or a ladder.
- Do not touch anyone or anything that is touching a downed electric wire.
- Never climb utility poles or play on fences surrounding electricity substations.

Lesson 3 Review

Assessment | Online Quiz

Visual Summary

An energy source produces an electric current in a circuit.

A series circuit has only one path for all devices in the circuit.

A parallel circuit has a separate path for each device in the circuit.

FOLDABLES

Use your lesson Foldable to review the lesson. Save your Foldable for the project at the end of the chapter.

What do you think NOW?

You first read the statements below at the beginning of the chapter.

5. In any electric circuit, current stops flowing in all parts of the circuit if a connecting wire is removed or cut.

6. The light energy given off by a flashlight comes from the flashlight's batteries.

Did you change your mind about whether you agree or disagree with the statements? Rewrite any false statements to make them true.

Use Vocabulary

1 Electric current decreases as more devices are added to a(n) _____ circuit.

2 Different amounts of current can flow through each device in a(n) _____ circuit.

Understand Key Concepts

3 One source of electric energy for an electric circuit is a
  A. battery.   C. switch.
  B. lightbulb. D. wire.

4 **List** the basic components of a simple electric circuit. Explain why each of the components is necessary.

5 **Contrast** How does the electric current change when more devices are added to a series circuit? To a parallel circuit?

Interpret Graphics

6 **Organize Information** Copy and fill in the graphic organizer below to show how electric energy is transformed to thermal energy by an electric current. Add additional boxes, if necessary.

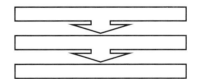

Critical Thinking

7 **Assess** A lightbulb and an electric motor are connected in a series circuit. Describe the amount of current in each device.

8 **Hypothesize** If a lightbulb in a circuit becomes dimmer, how did the energy supplied to the lightbulb change? Explain your answer.

9 **Create** Should a string of holiday lights be connected as a series circuit or as a parallel circuit? Create a graphic organizer to outline the advantages and disadvantages of the two types of circuits.

Lesson 3
EVALUATE

Inquiry Lab

2 class periods

Design an Elevator

Materials

thread, 1 m

washers

2 D-cell batteries in plastic bases

small hobby-type motor

2, 3-V bulbs in bases

alligator clip wires

cork

Also needed: paper clips, masking tape, stopwatch

Safety

There are many things an electric current can do besides making lightbulbs glow. One common electrical device is an electric motor. An electric motor uses an electric current to produce motion. Motors come in all sizes, from 300 times smaller than the diameter of a human hair to as large as a medium-sized house. We use motors for many different jobs—from driving a conveyor belt in a factory to operating a blender in the kitchen. One important use is to lift heavy objects.

Ask a Question
Where could it be important to be able to control the speed of an electric motor?

Make Observations

1. Read and complete a lab safety form.
2. Connect the thread to the cork on the motor's shaft. Attach a paper clip to the end of the thread. Fasten the motor to the edge of the table. Hang several washers on the paper clip over the edge of the table.
3. Draw your setup in your Science Journal.
4. Connect the battery, one lightbulb, and the motor in a series circuit. Observe whether the motor lifts the washers.
5. Adjust the number of washers until the motor lifts the washers as the thread winds up on the cork.
6. Record the time it takes for the motor to lift the washers.
7. Design a new circuit that will lift the washers at a different speed. Your circuit must include at least one battery and one lightbulb with the motor.

6

Lab Tips

☑ Secure the string tightly to the cork on the motor's drive shaft. You want the string to wind around the cork.

☑ Secure the motor firmly to the table so the weight does not pull it off.

☑ Think about the voltage used in your motor and the current through flowing in the motor. How can you increase both?

Form a Hypothesis

8 After observing the behavior of your circuits, formulate a hypothesis about what affects the speed of the motor.

Test Your Hypothesis

9 Design a circuit based on your hypothesis. The circuit should include at least one battery and one lightbulb. The circuit should lift the washers faster than your previous two circuits.

10 Create and test your new circuit. Record your observations.

Analyze and Conclude

11 Relate your hypothesis to the time each circuit took to lift the washers.

12 Compare your results with those of your classmates. Describe how the fastest time was achieved.

13 The Big Idea Explain how energy was transferred and transformed in your circuits.

Communicate Your Results

Create a pamphlet to sell your elevator design to a prospective buyer. Include a diagram and an explanation of how it works.

 **Extension**

Design a device that uses an electric motor to provide forward motion instead of vertical motion.

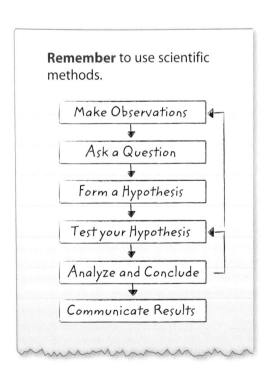

Remember to use scientific methods.

- Make Observations
- Ask a Question
- Form a Hypothesis
- Test your Hypothesis
- Analyze and Conclude
- Communicate Results

Chapter 19 Study Guide

 An energy source, such as a battery, creates an electric current in a circuit. Electrical devices in the circuit transform the electric energy of the current to useful forms of energy.

Key Concepts Summary

Lesson 1: Electric Charge and Electric Forces

- Particles that have the same type of electric charge repel each other. Particles that have different types of electric charge attract each other.
- Objects become negatively charged when they gain electrons, and become positively charged when they lose electrons.
- An **electric discharge** is the loss of **static charge.**

Lesson 2: Electric Current and Simple Circuits

- An **electric current** is the flow of electrically charged particles through a conductor.
- According to Ohm's law, across any portion of an **electric circuit, voltage** (V), current (I), and **electric resistance** (R) are related by the equation $V = IR$.

Lesson 3: Describing Circuits

- Electric circuits have a source of electric energy to produce an electric current, one or more electric devices to transform electric energy to useful forms of energy, and wires to connect the circuit's device(s) to the energy source.
- A **series circuit** has one path in which current flows. A **parallel circuit** has more than one path in which current flows..

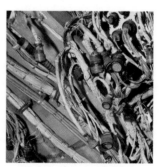

Vocabulary

static charge p. 680
electric insulator p. 682
electric conductor p. 682
polarized p. 683
electric discharge p. 685
grounding p. 685

electric current p. 690
electric circuit p. 690
electric resistance p. 692
voltage p. 693
Ohm's law p. 694

series circuit p. 701
parallel circuit p. 702

Study Guide

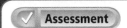

- Personal Tutor
- Vocabulary eGames
- Vocabulary eFlashcards

Chapter Project

Assemble your lesson Foldables as shown to make a Chapter Project. Use the project to review what you have learned in this chapter.

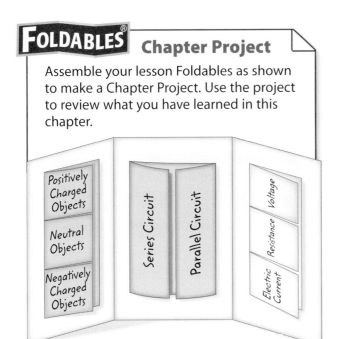

Use Vocabulary

1. An unbalanced electric charge is sometimes called a(n) _____.

2. Ohm's law states that as the voltage of a circuit increases, the current _____.

3. Electric charges flow easily in an electric _____.

4. A measure of the energy transformed in a portion of an electric circuit is _____.

5. Electric current decreases as more devices are added to a(n) _____ circuit.

6. Lightning is a(n) _____ that occurs in a fraction of a second.

7. A closed path in which electric charges flow is an electric _____.

8. Disconnecting one device in a _____ circuit will not cause other devices in the circuit to stop working.

Link Vocabulary and Key Concepts

Interactive Concept Map

Copy this concept map, and then use vocabulary terms from the previous page and other terms from the chapter to complete the concept map.

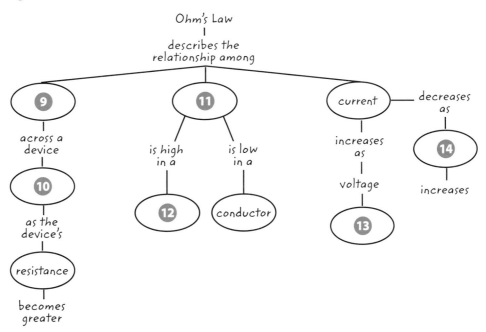

Chapter 19 Review

Understand Key Concepts

1. The measure of the energy transformed between two points in an electric circuit is
   A. resistance.
   B. voltage.
   C. electric current.
   D. electric force.

2. The switch in a circuit breaker opens when which of the following in the circuit becomes too high?
   A. current
   B. resistance
   C. static charge
   D. total charge

3. The electric force between two charges increases when
   A. both charges become negative.
   B. both charges become positive.
   C. the charges get closer together.
   D. the charges get farther apart.

4. Which best describes an electrically charged object?
   A. an electron without a static charge
   B. an object that has gained or lost electrons
   C. an object that has gained or lost neutrons
   D. a proton without a static charge

5. In the figure below, what form of energy in the battery is being converted to electric energy?

   A. chemical
   B. light
   C. nuclear
   D. thermal

6. What property of a wire changes when the wire is made thicker?
   A. current
   B. resistance
   C. static charge
   D. total electric charge

7. An electric field surrounds
   A. an electron.
   B. a neutron.
   C. both an electron and a neutron.
   D. neither an electron nor a neutron.

8. Which is a good conductor of electricity?
   A. glass
   B. gold
   C. plastic
   D. wood

9. Electric energy transforms into which forms of energy as electrons collide with atoms in a circuit?
   A. gravitational and light
   B. kinetic and nuclear
   C. thermal and light
   D. thermal and nuclear

10. Which lightbulbs in the diagram below will remain lit if the wire is disconnected at point B?

    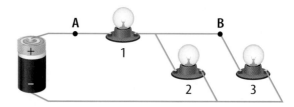

    A. 1 and 2
    B. 1 and 3
    C. 2 and 3
    D. 1, 2, and 3

Chapter Review

Assessment
Online Test Practice

Critical Thinking

11 Suggest What are two ways the electric force between two charged objects can be increased?

12 Evaluate Dry air is a better electric insulator than humid air. Would the electric discharge from a charged balloon happen more slowly in dry air or humid air? Explain your answer.

13 Recommend If a metal wire is made thinner, how would you change the length of the wire to keep the electric resistance the same?

14 Suggest What are two ways to increase electric current in a simple circuit?

15 Evaluate The current in a lightbulb stays the same, but the voltage across the lightbulb decreases. How does the electric energy to the lightbulb change?

16 Recommend Copy the diagram above and show on your drawing where a switch could be placed that would turn only lightbulb B off and on.

17 Assess Lightbulb A and lightbulb B are connected in a circuit. The voltage across lightbulb A is higher than the voltage across lightbulb B. Which lightbulb is brighter?

Writing in Science

18 Write a short essay describing some of the electrical devices you use every day. Include in your essay the energy transformations that occur in each device.

REVIEW THE BIG IDEA

19 How do electric charges flowing in a circuit regain some of the energy they transfer to atoms in a circuit?

20 The photograph below shows Nikola Tesla sitting and reading in his laboratory. How could he have used this picture to convince people that it was safe to use electricity in their homes?

Math Skills

Review
Math Practice

Use a Simple Equation

21 A current of 20 A flows into a hot water heater. If the electric resistance of the hot water heater is 12 Ω, what is the voltage across the hot water heater?

22 A refrigerator is plugged into an electrical outlet. If the voltage across the refrigerator is 120 V and the current flowing into the refrigerator is 6 A, what is the total electric resistance of the refrigerator?

23 What is the current in a flashlight bulb if the bulb has an electric resistance of 60 Ω and the voltage across the bulb is 6 V?

24 A window fan has a resistance of 80 Ω and the voltage across the fan is 120 V. How much current flows in the fan?

Chapter 19 Review • **711**

Standardized Test Practice

Record your answers on the answer sheet provided by your teacher or on a sheet of paper.

Multiple Choice

Use the figures to answer questions 1 and 2.

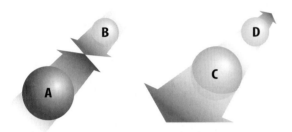

1 What must be true of particles A and B?
  A A and B must be like charges.
  B A and B must be opposite charges.
  C A must be positive, and B must be negative.
  D A must be negative, and B must be positive.

2 If the amount of charge on particles A, B C, and D are equal, what must be true of particle pairs AB and CD?
  A The attractive force of AB is greater than the repulsive force of CD.
  B The attractive force of AB is less than the repulsive force of CD.
  C The attractive force of AB equals the repulsive force of CD.
  D The repulsive force of AB is less than the attractive force of CD.

3 The movement of negative charges between nonconducting objects that are touching is called transferring charge by
  A conduction.
  B contact.
  C induction.
  D polarization.

4 Which phenomenon is an example of electric discharge?
  A friction
  B lightning
  C magnetism
  D static cling

5 The charged particles that move as an electric current are
  A electrons.
  B light particles.
  C metal atoms.
  D neutrons.

Use the table below to answer question 6.

Circuit	Current (A)	Resistance (Ω)
1	0.25	220
2	0.5	220
3	0.5	110

6 Based on the data, use Ohm's law to determine which statement is true.
  A Circuits 1 and 2 have the same voltage.
  B Circuits 2 and 3 have the same voltage.
  C Circuits 1 and 3 have the same voltage.
  D Each circuit has a different voltage.

7 When negative charges concentrate at one end of an object that is made of a conducting material, the object is
  A inducted.
  B insulated.
  C polarized.
  D undergoing friction.

Standardized Test Practice

Use the figure below to answer questions 8–10.

8 When the wires in this circuit are connected and all bulbs are lit, what type of circuit is shown?

- **A** negative
- **B** parallel
- **C** positive
- **D** series

9 If the broken wire of the top branch is connected, what will be true of the lightbulbs?

- **A** All of the bulbs would be lit.
- **B** None of the bulbs would be lit.
- **C** Only the top bulb would be lit.
- **D** Only the top two bulbs would be lit.

10 Connecting then disconnecting the broken ends of the top wire is similar to

- **A** adding then removing a fourth lightbulb.
- **B** adding then removing a second battery.
- **C** turning a switch on and off.
- **D** turning the middle lightbulb on and off.

Constructed Response

Use the figure below to answer questions 11 and 12.

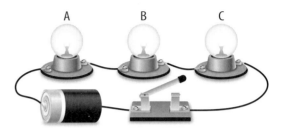

11 Identify the type of circuit shown. Explain what happens if the switch is closed. Explain what happens if the wire connecting lightbulbs A and B is removed.

12 Describe the kind of energy transformation(s) that takes place when the switch is closed.

13 When you turn off a lamp in your home, why do other electrical devices stay on? What does flipping the light switch do? What type of circuit does a house have?

14 What happens to the electric current in a circuit when the circuit's voltage is increased but resistance stays the same? What equation represents this relationship?

15 A string of 120 holiday lights is connected as a series circuit. The voltage across each bulb is 1 V. Another string of lights is connected as a parallel circuit. The voltage across each bulb is 120 V. Explain this voltage difference when both strings are plugged into a 120-V outlet.

NEED EXTRA HELP?															
If You Missed Question…	1	2	3	4	5	6	7	8	9	10	11	12	13	14	15
Go to Lesson…	1	1	1	1	2	2	1	2	2	2	2	3	2	3	3

Chapter 20
Magnetism

 How are electric charges and magnetic fields related?

 How loud is a magnet?

Since it was first manufactured in 1932, the electric guitar has inspired many new types of music. It is a prominent instrument in rock music and one of the most famous instruments to originate in the United States.

- How does an electric guitar use magnets to produce sound?
- Where are the magnets in the picture?
- How are electric charges and magnetic fields related?

Get Ready to Read

What do you think?

Before you read, decide if you agree or disagree with each of these statements. As you read this chapter, see if you change your mind about any of the statements.

1. All metal objects are attracted to a magnet.
2. Two magnets can attract or repel each other.
3. A magnetic field surrounds a moving electron.
4. Unlike a permanent magnet, an electromagnet has two north magnetic poles or two south magnetic poles.
5. A battery produces an electric current that reverses direction in a regular pattern.
6. An electric generator transforms thermal energy into electric energy.

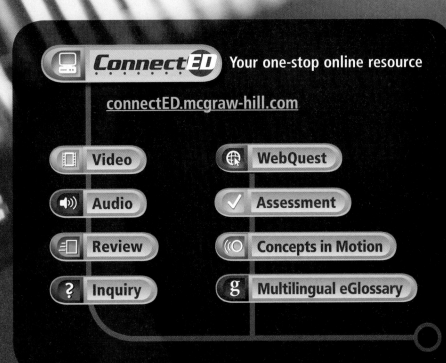

ConnectED — Your one-stop online resource

connectED.mcgraw-hill.com

- Video
- WebQuest
- Audio
- Assessment
- Review
- Concepts in Motion
- Inquiry
- Multilingual eGlossary

Lesson 1

Reading Guide

Key Concepts
ESSENTIAL QUESTIONS

- What types of forces do magnets apply to other magnets?
- Why are some materials magnetic?
- Why are some magnets temporary while others are permanent?

Vocabulary

magnet p. 717
magnetic pole p. 718
magnetic force p. 718
magnetic material p. 721
ferromagnetic element p. 721
magnetic domain p. 722
temporary magnet p. 723
permanent magnet p. 723

Multilingual eGlossary

Video BrainPOP®

Magnets and Magnetic Fields

Inquiry Is it a magnet?

If you ever played with magnets, you probably noticed that they seemed to have magical powers. Have you ever wondered just what it is that makes magnets so special and necessary for everyone on Earth?

Inquiry Launch Lab

10 minutes

What does *magnetic* mean?

Many years ago, people discovered that some minerals attract each other. The Greeks named these minerals *magnets*, after the ancient city Magnesia. What materials do magnets attract?

	Magnet "N" End	Magnet "S" End	Nail Point	Nail Head	Rubbed Nail Point	Rubbed Nail Head	Dropped Nail Point	Dropped Nail Head
Paper clip								
Paper								
Aluminum foil								
Choice #1								
Choice #2								

1. Read and complete a lab safety form.
2. Copy the data table into your Science Journal.
3. Touch each end of a **magnet** to a **paper clip, a piece of paper, a piece of aluminum foil,** and **two objects of your choosing.** Record your observations.
4. Repeat step 3, this time touching the ends of a **nail** to each object.
5. Rub the nail 25 times in the same direction across one end of the magnet. Repeat step 3, using the rubbed nail.
6. Drop the nail several times onto a hard surface. Repeat step 3 using the dropped nail. Record your observations.

Think About This

1. When does the nail behave like the magnet?
2. How do the two ends of the magnet interact with the various materials?
3. **Key Concept** What types of materials does a magnet attract?

Magnets

Did you use a computer or a hair dryer today? Did you listen to a stereo system? It might surprise you to know that all these devices contain magnets. Magnets also are used to produce the electric energy that makes these familiar devices work. *A **magnet** is any object that attracts the metal iron.*

As you just read, magnets are used in many ways. Therefore, magnets are manufactured in many shapes and sizes, as shown in **Figure 1.** You might be familiar with common bar magnets and horseshoe magnets. Some of the magnets holding papers on your refrigerator might be disc-shaped or flat and flexible. However, all magnets have certain things in common, regardless of their shape and size.

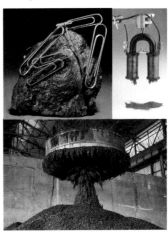

Figure 1 Magnets come in many sizes and shapes.

Lesson 1
EXPLORE

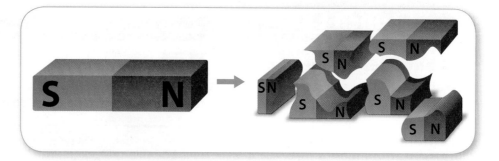

Figure 2 When a magnet is broken into pieces, each piece is still a magnet. ▶

Magnetic Poles

You might have noticed that the ends of some magnets are different colors or labeled *N* and *S*. These colors and labels identify the magnet's magnetic poles. *A* **magnetic pole** *is a place on a magnet where the force it applies is strongest.* There are two magnetic poles on all magnets—a north pole and a south pole. As shown in **Figure 2,** if you break a magnet into pieces, each piece will have a north pole and a south pole.

The Forces Between Magnetic Poles

A force exists between the poles of any two magnets. If similar poles of two magnets, such as north and north or south and south, are brought near each other, the magnets repel. This means that the magnets will push away from each other. If the north pole of one magnet is brought near the south pole of another magnet, the two magnets attract. This means the magnets will pull together. In other words, as shown in **Figure 3**, similar poles repel, and opposite poles attract.

A force of attraction or repulsion between the poles of two magnets is a **magnetic force.** A magnetic force becomes stronger as magnets move closer together and becomes weaker as the magnets move farther apart.

Key Concept Check What types of forces do magnets apply to other magnets?

WORD ORIGIN
pole
from Greek *polos,* means "pivot"

Make a vertical three-tab book, and label it as shown. Use it to organize your notes about magnetic forces.

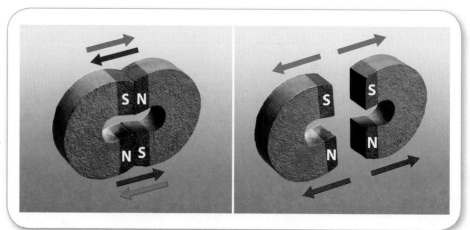

Figure 3 When the opposite poles of two magnets are close to each other, they attract. When similar poles of two magnets are close to each other, they repel. ▶

Concepts in Motion
Animation

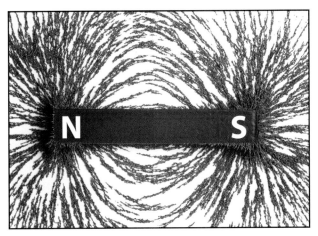 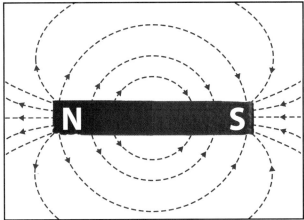

▲ **Figure 4** The magnetic field around a magnet can be shown with iron filings. Drawings of magnetic field lines include arrows to show the north-to-south direction of the magnetic field.

Magnetic Fields

Recall from a previous chapter that charged objects repel or attract each other even when they are not touching. Similarly, magnets can repel or attract each other even when they are not touching. An invisible magnetic field surrounds all magnets. It is this magnetic field that applies forces on other magnets.

Magnetic Field Lines

In **Figure 4,** iron filings have been sprinkled around a bar magnet. The iron filings form a pattern of curved lines that reveal the magnet's magnetic field.

A magnet's magnetic field can be represented by lines, called magnetic field lines. Magnetic field lines have a direction, that is shown on the right in **Figure 4.** Notice that the lines are closest together at the magnet's poles. This is where the magnetic force is strongest. As the field lines become farther apart, the field and the force become weaker.

Combining Magnetic Fields

What happens to the magnetic fields around two bar magnets that are brought together? The two fields combine and form one new magnetic field. The pattern of the new magnetic field lines depends on whether two like poles or two unlike poles are near each other, as shown in **Figure 5.**

SCIENCE USE V. COMMON USE

field

Science Use a space in which a given effect (such as magnetism) exists

Common Use an open area free of woods and buildings

Figure 5 When magnetic fields combine, they form field lines with different patterns. The new patterns depend on whether the magnets are attracting or repelling each other. ▼

Attraction

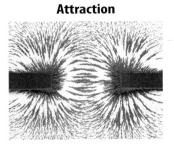

When opposite poles of two magnets are near each other, the resulting magnetic field applies a force of attraction.

Repulsion

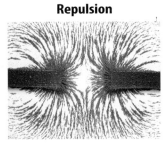

When like poles of two magnets are near each other, the resulting magnetic field applies a force of repulsion.

Lesson 1
EXPLAIN

Earth's Magnetic Field

A magnetic field surrounds Earth similar to the way a magnetic field surrounds a bar magnet. Earth has a magnetic field due to molten iron and nickel in its outer core. Like all magnets, Earth has north and south magnetic poles.

Compasses

Have you ever wondered why a compass needle points toward north? The needle of a compass is a small magnet. Like other magnets, a compass needle has a north pole and a south pole. Earth's magnetic field exerts a force on the needle, causing it to rotate. **Figure 6** shows that if a compass needle is within any magnetic field, including Earth's, it will line up with the magnet's field lines. A compass needle does not point directly toward the poles of a magnet. Instead, the needle aligns with the field lines and points in the direction of the field lines. Earth's magnetic poles and geographic poles are not in the same spot, as shown in **Figure 7**. Therefore, you cannot find your way to the geographic poles with only a compass.

Auroras

Earth's magnetic field protects Earth from charged particles from by the Sun. These particles can damage living organisms if they reach the surface of Earth. Earth's magnetic field deflects most of these particles. Sometimes, large numbers of particles from the Sun travel along Earth's magnetic field lines and concentrate near the magnetic poles. There, the particles collide with atoms of gases in the atmosphere, causing the atmosphere to glow. The light forms shimmering sheets of color known as auroras. An aurora is shown in **Figure 7**.

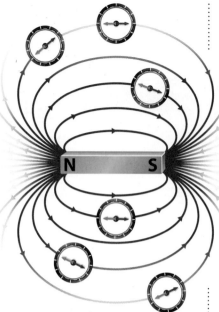

▲ **Figure 6** A compass needle will align with a magnet's magnetic field lines.

Visual Check Why do all the compass needles in **Figure 6** point in different directions?

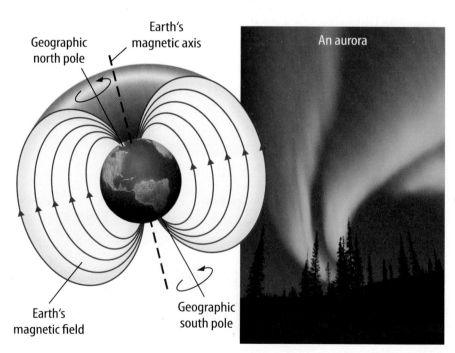

Figure 7 A magnetic field surrounds Earth. This magnetic field can cause auroras to occur near Earth's magnetic north pole and south pole. ▶

Magnetic Materials

You have read that magnets attract other magnets. But, magnets also attract objects, such as nails, which are not magnets. *Any material that is strongly attracted to a magnet is a **magnetic material**.* Magnetic materials often contain ferromagnetic (fer oh mag NEH tik) elements. **Ferromagnetic elements** *are elements, including iron, nickel, and cobalt, that have an especially strong attraction to magnets.*

It is important to understand that while not all materials are magnetic, a magnetic field does surround every atom that makes up all materials. The field is created by the atom's constantly moving electrons. The interactions of these individual fields determines whether a material is magnetic. Next, you will read about why magnetic materials make good magnets.

 Reading Check Why is nickel a magnetic material?

> **REVIEW VOCABULARY**
> **element**
> a substance made up of only one kind of atom

Inquiry MiniLab 20 minutes

Where is magnetic north?

Many animals such as birds use Earth's magnetic field for navigation. With the help of a compass, people can find their way, too. What does it mean when a compass points north?

1. Read and complete a lab safety form.
2. On a piece of **paper**, draw a circle with a diameter 2 cm longer than a **bar magnet**. Place the bar magnet in the center of your circle. Label the magnet's north and south poles on the circle.
3. Place a **compass** on the paper just outside the circle. Draw an arrow pointing in the same direction as the north end of the compass.
4. Repeat step 3 in at least eight different places around the circle.
5. In your Science Journal, indicate to which pole of the bar magnet the north end of the compass needle points.
6. Now, step away from your magnet and all metal objects in the room. Hold the compass level in your hand. Face the direction indicated by the compass. Record the direction you are facing in your Science Journal.

Analyze and Conclude

1. **Describe** the magnetic pole that attracts the north end of a compass needle.
2. **Key Concept** Hypothesize about the interaction of Earth's magnetic field and the compass needle.

Magnetic Domains

If all atoms act like tiny magnets, why do magnets not attract most materials, such as glass and plastic? The answer is that in most materials, the magnetic fields of the atoms point in different directions, as in part a of **Figure 8**. As the fields around the atoms combine, they cancel each other. Thus, most types of matter are nonmagnetic materials and are not attracted to magnets.

In a magnetic material, atoms form groups called magnetic domains. *A **magnetic domain** is a region in a magnetic material in which the magnetic fields of the atoms all point in the same direction*, as shown in part b of **Figure 8**. The magnetic fields of the atoms in a domain combine, forming a single field around the domain. Each domain is a tiny magnet with a north pole and a south pole.

When Domains Don't Line Up

Many magnetic materials are not magnets. This is because the magnetic fields of their domains point in random directions, as shown in part b of **Figure 8**. Similar to the atoms of a nonmagnetic material, the random magnetic fields of the domains cancel each other. Even though the individual domains are magnets, the entire object has no effective magnetic field.

When Domains Line Up

How do a bar magnet and a steel nail that is not a magnet differ? Both are magnetic materials. Both have atoms grouped into magnetic domains. However, for an object to be a magnet, its magnetic domains must align as in part c of **Figure 8**. When the domains align, their magnetic fields combine, forming a single magnetic field around the entire material. This causes the object to become a magnet.

Key Concept Check Why are some materials magnetic?

Academic Vocabulary

align
(verb) to arrange in a straight line; adjust according to a line

Figure 8 In many materials, atoms act like tiny magnets. Most materials are nonmagnetic materials (a), but some are magnetic with magnetic domains (b). In a magnet, the poles of the domains line up (c).

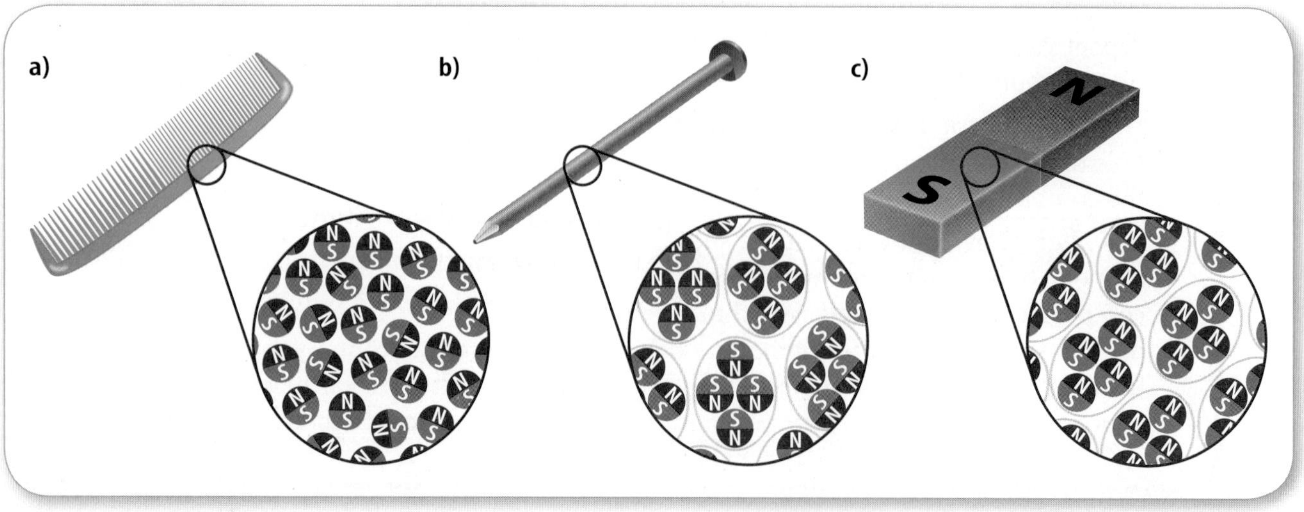

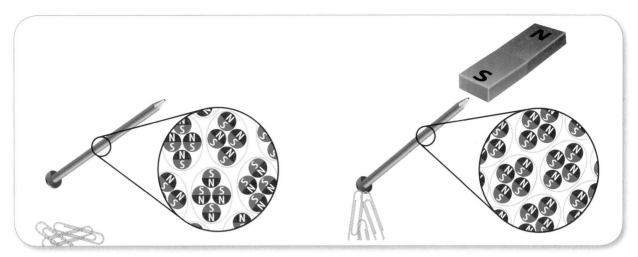

How Magnets Attract Magnetic Materials

Figure 9 shows a nail coming close to one of the poles of a bar magnet. Remember, even though the nail itself is not a magnet, each magnetic domain in the nail is a small magnet. The magnetic field around the bar magnet applies a force to each of the nail's magnetic domains. This force causes the domains in the nail to align along the bar magnet's magnetic field lines. As the poles of the domains of the nail point in the same direction, the nail becomes a magnet. Now the nail can attract other magnetic materials, such as paper clips.

Temporary Magnets

In **Figure 9,** the nail becomes a temporary magnet. *A magnet that quickly loses its magnetic field after being removed from a magnetic field is a* **temporary magnet.** The nail is a magnet only when it is close to the bar magnet. There, the magnetic field of the bar magnet is strong enough to cause the nail's magnetic domains to line up. However, when you move the nail away from the bar magnet, the poles of the domains in the nail return to pointing in different directions. The nail no longer is a magnet and no longer attracts other magnetic materials.

 Reading Check Why would the nail in **Figure 9** lose its magnetic field if it is moved away from the bar magnet?

Permanent Magnets

A magnet that remains a magnet after being removed from another magnetic field is a **permanent magnet.** In a permanent magnet, the magnetic domains remain lined up. Some magnetic materials can be made into permanent magnets by placing them in a very strong magnetic field. This causes the magnetic domains to align and stay aligned. The material then remains a magnet after it is removed from the field.

 Key Concept Check Why are some magnets temporary while others are permanent?

Figure 9 A nail is made of a magnetic material. It becomes a temporary magnet when it is close to another magnet.

Visual Check What happens to the magnetic domains of the nail as it becomes a temporary magnet?

Lesson 1 Review

Visual Summary

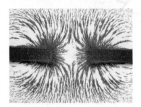

When the magnetic fields of two magnets combine, a magnetic force exists between the magnets.

Objects made of magnetic materials are attracted to magnets.

The magnetic domains of a permanent magnet are aligned even when the magnet is not near another magnetic field.

FOLDABLES

Use your lesson Foldable to review the lesson. Save your Foldable for the project at the end of the chapter.

What do you think NOW?

You first read the statements below at the beginning of the chapter.

1. All metal objects are attracted to a magnet.

2. Two magnets can attract or repel each other.

Did you change your mind about whether you agree or disagree with the statements? Rewrite any false statements to make them true.

Use Vocabulary

1 Elements that are strongly attracted to a magnet are _____.

2 The magnetic poles of the atoms in a(n) _____ point in the same direction.

3 An object that attracts other objects made of magnetic materials is a(n) _____.

Understand Key Concepts

4 Identify Where around a magnet is the magnetic field strongest?

5 Which does NOT contain magnetic domains?
   A. bar magnet   C. paper airplane
   B. disc magnet D. steel nail

6 Contrast the magnetic domains in a steel nail that is far from a bar magnet and one that is close to a bar magnet.

Interpret Graphics

7 Organize Copy and fill in the table below to describe the type of force between the magnetic poles.

Poles	Type of Force
South pole and south pole	
North pole and north pole	
South pole and north pole	

Critical Thinking

8 Predict Where would the north pole of a compass needle point if Earth's magnetic field switched direction?

9 Evaluate Two nails have identical sizes and shapes. In one nail, 20 percent of the domains are lined up. In the other nail, 80 percent of the domains are lined up. Which has a stronger magnetic field?

Roller Coasters

SCIENCE & SOCIETY

How Magnets Make a Wild Ride a Safe Ride

Roller coasters certainly have changed throughout their long history! Four hundred years ago, the ancestors of roller coasters appeared in Russia. These rides were long, steep wooden slides covered in ice. Some were over 20 m high. Riders slid down the slope in sleds made of wood or blocks of ice, crashing into a sand pile. Today's roller coasters can reach speeds over 160 km/h! Crashing into a sand pile at this speed is no way to stop a ride safely!

Due to their safety, magnetic brakes are a technology that is gaining popularity with roller-coaster designers. Magnetic brakes rely on the interaction of magnetic fields—these brakes never come in contact with the cars. One design of magnetic brake uses rows of strong magnets built at the side of the track. A metal fin on the car passing between the rows of magnets creates a magnetic field that opposes the fin's motion. Magnetic braking is virtually fail-safe because it relies on the basic properties of magnetism and requires no electricity.

Engineers are designing future roller coasters with magnets to hold the cars to the rails during death-defying drops at extreme speeds never before possible. Computerized sensors and automatic detectors on new generation roller coasters will control the entire roller coaster experience from start to finish. Will science and technology soon make having fun on a roller coaster a process as sophisticated as a rocket launch?

Metal fin: typically copper or a copper/aluminum alloy.

Neodymium magnets: an alloy of neodymium, iron, and boron, currently the strongest type of permanent magnet.

Hydraulic system: High-pressure fluids push against a piston to lower the magnets when braking is not needed.

It's Your Turn

RESEARCH AND REPORT Roller-coaster cars are not the only kind of vehicles that make use of magnets. Research maglev trains to find out more about how these trains use magnets and magnetic fields to provide a safe, smooth ride. Write a short report to share what you learn.

Lesson 1 EXTEND

Lesson 2

Reading Guide

Key Concepts 🔑
ESSENTIAL QUESTIONS

- Why does a magnet apply a force on an electric current?
- How do electromagnets and permanent magnets differ?
- How do electric motors use magnets?

Vocabulary
electromagnet p. 728
electric motor p. 730

g Multilingual eGlossary

Making Magnets with an Electric Current

Inquiry What can magnets find?

What does *metal detector* mean to you? Do you think of searching for buried treasure? Are you reminded of airport security or of the hand-held scanners at a sporting event? Metal-detector technology is a huge part of our lives. How are electricity and magnetism used in these devices?

Inquiry Launch Lab

20 minutes

When is a wire a magnet?

In 1820, Hans Øersted made a remarkable discovery about how magnetism and electricity are parts of the same thing. How do magnetism and electricity relate?

1. Read and complete a lab safety form.
2. With a **clamp attachment,** hang a length of **insulated wire** through the center of an iron ring on a **ring stand.** Make a small hole in the center of a 10-cm square piece of **cardboard.** Slide the wire through the hole. Rest the card on the iron ring. Secure the cardboard to the iron ring with **tape.**
3. Place four **small compasses** on the card in a circle around the wire.
4. In your Science Journal, draw your setup. Include the wire through the card and the circle of compasses in your drawing. Indicate the direction that each compass points.
5. Use **alligator clip wires** to connect two **D-cell batteries in holders** in series with the ends of the wire. Draw a second diagram. Show the direction the compasses point with the batteries connected.
   ⚠ Unhook the wire after a few seconds to prevent it from overheating!

Think About This

1. What would happen if you reversed the batteries?
2. **Key Concept** Why do you think the compass needles behaved as they did after the batteries were connected?

Moving Charges and Magnetic Fields

Recall that a magnetic field surrounds a magnet. In addition, a magnetic field surrounds an electric current. This is why a compass needle moves when placed near a current-carrying wire. The needle moves because the magnetic field around the wire applies a force to the compass needle.

The Magnetic Field Around a Current

A magnetic field surrounds all moving charged particles. Remember that an electric current is the flow of electric charge. In a current-carrying wire, the magnetic fields of the flowing charges combine to produce a magnetic field around the wire, as shown in **Figure 10.** The field around the wire becomes stronger as the current in the wire increases, or as more electrons flow in the wire.

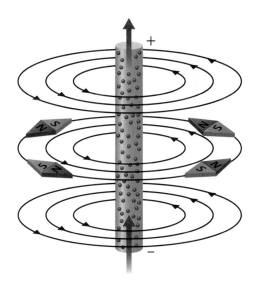

Figure 10 The magnetic field around a current-carrying wire forms closed circles.

✓ **Visual Check** Why do compass needles around a current-carrying wire point in different directions?

▲ Figure 11 A magnet applies a force to a current-carrying wire. The direction of the force is related to the direction of the current.

Review Personal Tutor

Magnets and Electric Currents

What happens when a magnet comes near a current-carrying wire? Because negatively charged electrons are moving in the wire, the magnet applies a force to the moving charges. This force causes the wire to move, as shown in **Figure 11**. The direction of the force on the wire depends on the direction of the current and the direction of the magnet's field.

Key Concept Check Why does a magnet apply a force to an electric current?

Electromagnets

What do a stereo speaker and a hair dryer have in common? Both use electromagnets. An **electromagnet** is *a magnet created by wrapping a current-carrying wire around a ferromagnetic core.* The core becomes a magnet when an electric current flows through the coil. If a device uses electricity and has moving parts, chances are an electromagnet is inside.

Reading Check What is an electromagnet?

Making an Electromagnet

Recall that an electric current in a wire produces a magnetic field around the wire. The strength of the magnetic field around a current-carrying wire increases when the wire is wound into a coil, as shown in the center of **Figure 12**. The magnetic fields around the individual loops of the coil combine, making the magnetic field around the wire stronger.

An electromagnet is made by placing a ferromagnetic material, such as iron, as a core within the wire coil, as shown on the right in **Figure 12**. The magnetic field of the coil causes the ferromagnetic core to become a magnet. The core greatly increases the strength of the coil's magnetic field. This makes the electromagnet's force stronger. And, as the number of loops in the coil increases, the magnetic field of the electromagnet becomes stronger.

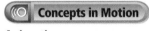
Animation

Figure 12 A ferromagnetic core becomes an electromagnet when surrounded by a current-carrying wire. ▼

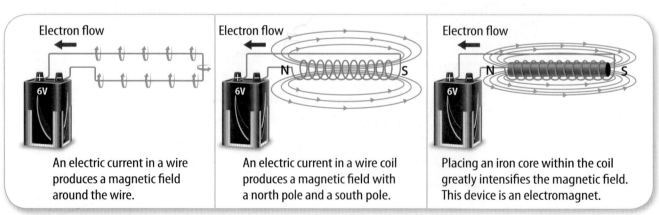

An electric current in a wire produces a magnetic field around the wire.

An electric current in a wire coil produces a magnetic field with a north pole and a south pole.

Placing an iron core within the coil greatly intensifies the magnetic field. This device is an electromagnet.

Properties of Electromagnets

An electromagnet is a special type of temporary magnet. An electromagnet differs from other magnets in several ways. First, the magnetic field around an electromagnet can be turned off by turning off the current. Without an electric current in the coil, an electromagnet is just a piece of iron wrapped with wire. But with the current on, it becomes a magnet.

Second, the strength of an electromagnet can be controlled. You read that the strength of the magnetic force of an electromagnet depends on the number of loops in the electromagnet's wire coil. Also, increasing the electric current in the coil strengthens the magnetic force.

Finally, the poles of an electromagnet can be reversed. Changing the direction of the current in the wire coil reverses an electromagnet's north and south magnetic poles. You will read more about this property in the next part of the lesson.

 Key Concept Check How do electromagnets and permanent magnets differ?

Using Electromagnets

Because an electromagnet has reversible magnetic poles, other magnets can attract and then repel an electromagnet. For example, a loudspeaker produces sound waves when the direction of the electric current in its electromagnetic coil rapidly and repeatedly reverses. **Figure 13** shows the electromagnet in a loudspeaker that connects to a paper or plastic drive cone. The changing direction of the current causes the electromagnet to attract and then repel a permanent magnet. This makes the cone vibrate, producing sound waves.

FOLDABLES Make a vertical three-tab Venn book. Use it to compare and contrast the properties of permanent magnets and electromagnets.

Figure 13 A rapidly changing electric current in the coil of an electromagnet causes the drive cone of a loudspeaker to produce sound waves.

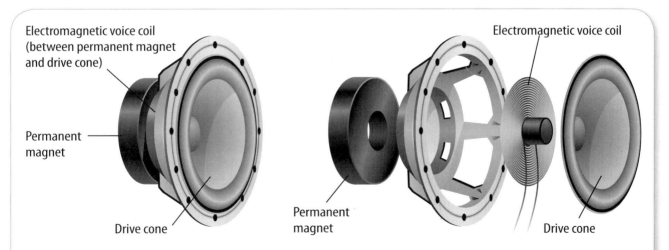

As the electric current in the voice coil changes direction (up to thousands of times per second), the coil's magnetic field changes with the current. The interaction of the coil's changing magnetic field and the permanent magnet causes the drive cone to move with a back-and-forth motion. The vibrating cone produces sound waves in the surrounding air.

Inquiry MiniLab

20 minutes

What is an electromagnet?

Electromagnets do things that permanent magnets cannot—they can be turned off and their strength altered. As a result, electromagnets are in many modern electrical devices.

1. Read and complete a lab safety form.
2. With **sandpaper,** rub off 2 cm of insulation from both ends of 150 cm of **magnet wire.**

3. Make a coil by wrapping half of the wire around half of a **drinking straw.** Leave a 5-cm tail of wire as you begin wrapping the wire. Approximately 75 cm of wire will remain at the other end of the coil.
4. Count how many **paper clips** the coil picks up and how it interacts with a **permanent magnet.** Record your observations in your Science Journal.
5. Use **alligator clip wires** to connect the tails of the coil to a **D-cell battery in a base.** Repeat step 4. Record your observations. ⚠ *Unhook the wire after a few seconds to avoid overheating!*
6. Slide a **nail** into the center of the straw. Repeat step 4. Record your observations.
7. Disconnect the battery. Remove the nail from the straw. Wrap the rest of the wire around the straw. Repeat steps 4 through 6. Record your observations.

Analyze and Conclude

1. **Explain** your observations of the paper clips lifted by the coil in steps 4, 5, 6, and 7.
2. 🔑 **Key Concept** Design a graphic organizer that shows possible uses of permanent magnets and electromagnets.

Magnets and Electric Motors

Power tools, electric fans, hair dryers, computers, and even microwave ovens use electric motors. *An **electric motor** is a device that uses an electric current to produce motion.* How many devices can you think of that use electric motors?

A Simple Electric Motor

A simple electric motor is shown in **Figure 14.** The main parts of an electric motor are a coil of wire connected to a rotating shaft, a permanent magnet, and a source of electric energy, such as a battery.

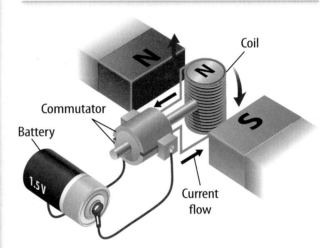

Figure 14 A rotating electromagnet in an electric motor is mounted between the poles of a permanent magnet. Here, a battery supplies a current to the electromagnet.

 **Concepts in Motion** Animation

Making the Motor Spin

Recall that one useful property of electromagnets is that their magnetic poles easily can be reversed. This property of electromagnets is what makes an electric motor spin.

Unlike Poles Attract When an electric current is supplied to the motor, the unlike poles of the permanent magnet and electromagnet attract each other, causing the motor to begin to turn.

Reversing the Electric Current As the unlike poles of the motor's electromagnet and permanent magnet line up, the attraction between them will stop the motor's spinning. To keep the motor turning, the poles of the electromagnet must reverse.

The poles reverse when the direction of the current in the electromagnet changes, as shown in **Figure 15.** The commutator is the device in the motor that reverses the direction of the current in the electromagnet.

Similar Poles Repel Now, the similar poles of the electromagnet and the permanent magnet are close to each other. Thus, the poles repel each other and the electromagnet keeps spinning.

To keep the electromagnet spinning, the direction of the current in the coiled wire must continue changing direction. The commutator reverses the current when the poles of the electromagnet come near the poles of the permanent magnet. The permanent magnet and the electromagnet continuously attract then repel each other. This makes the electromagnet continue to spin.

Using Electric Motors

You read that an electric motor includes an electromagnet mounted on a shaft. As the electromagnet spins, the shaft spins, too. This spinning motion can be used to create other motions. Can you think of any parts powered with an electric motor in a car?

A system of levers connected to an electric motor produces the back-and-forth motion of the windshield wipers. Electric motors make the power windows go up and down. A compact disk player uses an electric motor to spin a CD and move a small laser across the disk. And now, electric motors are replacing gasoline engines in many cars.

Key Concept Check How are magnets used in electric motors?

Figure 15 The controlled forces of attraction and repulsion between an electromagnet and a permanent magnet make an electric motor spin.

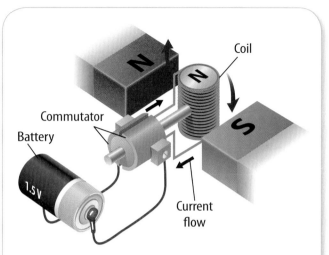

The opposite magnetic poles of the electromagnet and the permanent magnet attract each other and make the electromagnet rotate.

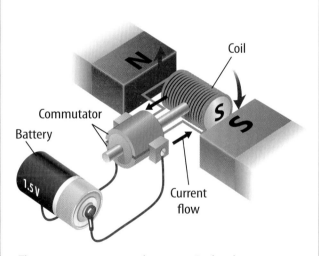

The commutator causes the current in the electromagnet to change direction, which reverses the poles of the electromagnet. The magnets now repel each other and the motor keeps rotating.

Visual Check What happens to the magnetic poles of the permanent magnets in an electric motor as the electromagnet rotates?

WORD ORIGIN
motor
from Latin *movēre*, means "to move"

Lesson 2 Review

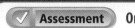 Assessment   Online Quiz

Visual Summary

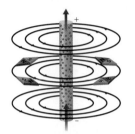

A current-carrying wire is surrounded by a magnetic field.

A magnetic core within a curent-carrying wire coil is an electromagnet.

Alternating attraction and repulsion between a permanent magnet and an electromagnet causes an electric motor to rotate.

FOLDABLES

Use your lesson Foldable to review the lesson. Save your Foldable for the project at the end of the chapter.

What do you think NOW?

You first read the statements below at the beginning of the chapter.

3. A magnetic field surrounds a moving electron.

4. Unlike a permanent magnet, an electromagnet has two north magnetic poles or two south magnetic poles.

Did you change your mind about whether you agree or disagree with the statements? Rewrite any false statements to make them true.

Use Vocabulary

1 A current-carrying wire coil around a ferromagnetic core is a(n) _____.

2 An electric current causes a shaft in a(n) _____ to rotate.

Understand Key Concepts

3 Identify Why does a magnet apply a force to a current-carrying wire?

4 Summarize How do electromagnets and permanent magnets differ?

5 In an electric motor, a force is applied to the electromagnet by
   A. a battery.
   B. a commutator.
   C. an electric current.
   D. a permanent magnet.

Interpret Graphics

6 Compare How would the magnetic field around the wire at the right change if the electrons flowed in the opposite direction?

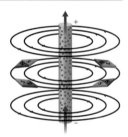

7 Sequence Events Copy and fill in the graphic organizer below to show the sequence of events that occur during one revolution of an electric motor.

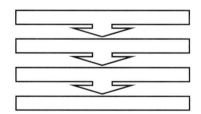

Critical Thinking

8 Predict A current-carrying wire made of nonmagnetic copper is attracted to a bar magnet. How will the force between the wire and the magnet change if more current flows in the wire?

732 Chapter 20
EVALUATE

Inquiry Skill Practice: Measure

40 minutes

How can you measure an electric current?

Materials

D-cell battery in a plastic holder

3 lightbulbs in screw bases

compass

sandpaper

straws

Also needed:
1.5 m of magnet wire, alligator clip wires

Safety

Moving electric charges create a magnetic field. A compass is a device that shows the direction of a magnetic field. How can you use a magnetic compass to measure an electric current?

Learn It

You **make measurements** every day. Scientists also make measurements. Sometimes the tools they use are familiar, such as rulers or scales. However, researchers must sometimes invent measuring tools based on their specific needs. What could you use to construct your own ammeter—a device that measures electric current?

Try It

1. Read and complete a lab safety form.
2. Copy the data table below into your Science Journal.

Number of Lightbulbs in Circuit	Position of Compass Needle
0	(needle shown deflected)
1	
2	
3	

3. Using sandpaper, strip 1 cm of insulation from both ends of 1.5 m of magnet wire.
4. Wrap the wire around a drinking straw. Do not crush the straw. The coils should be close together and near one end of the straw. Leave 10 cm of wire sticking out from both sides.
5. Tape the compass and the straw on a table so the straw is perpendicular to the compass needle.
6. Using a battery, a lightbulb, and your coil, create a closed-series circuit. Observe the behavior of the compass. Record your observations in your data table. ⚠ *Unhook the wire after a few seconds to avoid overheating!*
7. Add a second lightbulb in series with the first. Record your observations of the compass needle.
8. Add a third lightbulb to your series circuit. Record your observations.

Apply It

9. **Explain** the relationship between the number of lightbulbs and the current.
10. **Key Concept** Explain how the compass measured the amount of an electric current.

Lesson 2
EXTEND
733

Lesson 3

Making an Electric Current with Magnets

Reading Guide

Key Concepts
ESSENTIAL QUESTIONS

- How can a wire and a magnet produce an electric current?
- How do electric generators create an electric current?
- How are transformers used to bring an electric current into your home?

Vocabulary

electric generator p. 736
direct current p. 737
alternating current p. 737
turbine p. 738
transformer p. 739

g Multilingual eGlossary

Inquiry How bright is a magnet?

Riding a bicycle safely at night means being visible to other vehicles. A small generator mounted against the bicycle's wheel provides the electric current that powers the lights, which help keep your ride safe. How does this simple device change the motion of your bicycle into a bright light at night?

Inquiry Launch Lab
20 minutes

Why does the pendulum slow down?
A pendulum swings back and forth. How can magnetism be used to stop a pendulum?

1. Read and complete a lab safety form.
2. Using a **thread**, suspend a **strong magnet** from a **ring stand.** Adjust the height of the ring so that the magnet hangs about 0.5 cm off the table.
3. Pull the pendulum back 5 cm, and gently release it. Count the swings until the pendulum comes to a stop. Record your result in your Science Journal.
4. Fold a **foil square** into a smooth 3-cm-wide strip. With **tape**, secure the foil under the pendulum, parallel to the direction of swing. The magnet should not touch the foil.
5. Repeat step 3. Record your results.

Think About This
1. How might your results differ if you used a marble or a baseball as the pendulum?
2. **Key Concept** How would you describe the interaction of the foil and the magnet?

Magnets and Wire Loops

In the previous lesson, you read that an electric current is surrounded by a magnetic field. You experience a magnetic field surrounding an electric current every time you use a device that has an electric motor. Now, you will read about how magnetic fields can produce electric currents. Electric power plants use powerful magnetic fields to produce the electric energy supplied to homes and businesses.

Generating Electric Current

How can a magnetic field produce an electric current? **Figure 16** shows one way. A magnet is moved through a wire coil that is part of a closed electric circuit. As the magnetic field moves over the wire coil, an electric current is produced in the circuit. When the magnet stops moving, there is no current in the circuit. The direction of the current depends on the direction in which the magnet moves.

Key Concept Check How can a wire and a magnet produce an electric current?

Figure 16 An electric current is produced in a wire coil when a magnet moves within the coil.

Visual Check If the magnet does not touch the wire coil, what causes a current to flow in the wire?

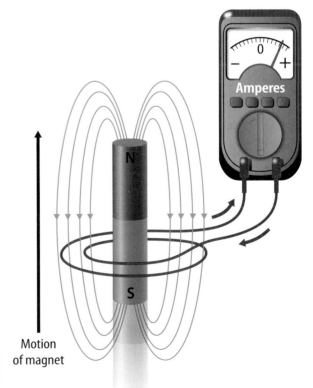

Motion of magnet

Lesson 3
EXPLORE
735

Figure 17 When a wire moves through a magnetic field, an electric current is produced in the circuit. There is no current when the coil is not moving.

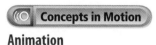

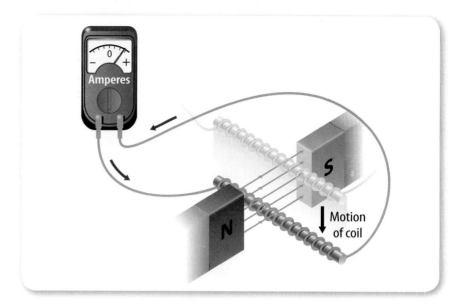

A Moving Wire and an Electric Current.

You just read that a magnetic field passing over a wire coil produces an electric current. **Figure 17** shows moving a wire coil through a magnetic field as another way to generate an electric current. The magnetic force between the magnet and the electrons in the wire causes the electrons in the wire to move as an electric current in the wire.

Either a magnetic field passing over a wire coil or a wire coil passing through a magnetic field produces an electric current. You will now read how this relationship powers the many electric devices you use everyday.

 Reading Check How can a magnetic field produce an electric current in a wire coil?

Electric Generators

When you turn on a flashlight, batteries produce an electric current in the lightbulb. However, when you turn on a TV, large electric generators in distant power plants provide the electric current to the TV. An **electric generator** is *a device that uses a magnetic field to transform mechanical energy to electric energy.*

A Simple Electric Generator

Figure 18 on the next page shows a hand generator in a circuit. The crank rotates a wire coil through the magnetic field of a small permanent magnet. This produces an electric current in the circuit. The current continues as the crank continues to rotate the coil within the magnetic field.

Unlike the hand-cranked generator that uses the magnetic field of a small permanent magnet, larger generators often use powerful electromagnets to produce a magnetic field.

Make a vertical three-tab book. Label it as shown. Use it to organize notes about the electric current in each device.

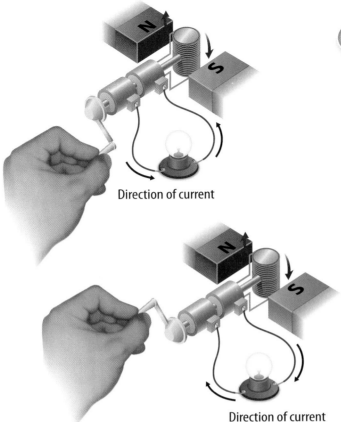

Direction of current

Direction of current

Figure 18 A generator produces an electric current when a wire coil rotates between the poles of two magnets.

Direct Current and Alternating Current

The electric current produced by a battery differs from the current produced by the generator in **Figure 18**. The current produced by a battery flows in a circuit in only one direction. *An electric current that flows in one direction is* **direct current.**

The current produced by a generator changes direction as the poles of the rotating coil line up with the poles of the magnet. *An electric current that changes direction in a regular pattern is* **alternating current.**

Some generators do produce direct current. They use a commutator, such as the one in **Figure 15** in Lesson 2. A commutator is a switch that rotates with the wire coil and prevents the current from changing directions. This results in direct current.

Key Concept Check How do generators produce electric current?

Inquiry MiniLab
20 minutes

How many paper clips can you lift?

With a hand generator, you provide the mechanical energy that causes a magnet to spin inside a wire coil. How does the speed of the spinning affect the output of a generator?

1. Read and complete a lab safety form.
2. Copy the table into your Science Journal.

Rate of Rotation	Number of Paper Clips	Rate of Rotation (opposite direction)	Number of Paper Clips
Not rotating		Not rotating	
Slow		Slow	
Fast		Fast	

3. Make an electromagnet by wrapping 150 cm of **magnet wire** around a **straw.** Leave 10-cm tails of wire at both ends of the coil.
4. Use **sandpaper** to remove 2 cm of insulation from each end of the wire. Connect the ends to a **hand generator.**
5. Slowly turn the crank of the generator.
6. Have a partner use the electromagnet to pick up **paper clips.** Record how many paper clips the electromagnet picks up.
7. Repeat step 6, turning the generator at a higher speed.
8. Hold a **compass** near one end of the coil. Turn the generator at different speeds. Record your observations.
9. Repeat steps 5–8, turning the generator crank in the opposite direction.

Analyze and Conclude

1. **Explain** how the speed of the generator and the direction it was turned affected your results.
2. **Key Concept** How does the movement of your hand lead to lifting the paper clips?

Lesson 3
EXPLAIN

▲ **Figure 19** Enormous electric generators at Hoover Dam on the Colorado River generate the electric energy used by more than 1 million households in the region.

WORD ORIGIN
turbine
from Latin *turban*, means "confusion"

Figure 20 This turbine will connect to an electric generator. There, it will spin as steam is forced through its blades. ▼

Generators and Power Plants

Electric power plants use huge generators, such as the ones shown in **Figure 19**. They generate the electric current used in thousands of homes. Instead of one wire coil, these generators have several large coils of wire. Each coil might have thousands of loops. Increasing the number of coils and the number of loops in each coil increases the amount of current a generator produces.

Mechanical Energy to Electric Energy

A useful energy transformation occurs in a generator. A generator produces an electric current as its wire coils rotate through a magnetic field. The rotating coils transform mechanical energy to electric energy. The electric energy is the kinetic energy of the current of electrons in the coil.

Supplying Mechanical Energy

As you turn a hand generator, you supply mechanical energy to the generator. The generators in electric power plants also need a source of mechanical energy to keep the coils rotating. Some power plants use the mechanical energy of high-pressure jets of steam from boiling water. Hydroelectric power plants use the mechanical energy of falling water.

In electric power plants, a generator usually is connected to a turbine (TUR bine), as shown in **Figure 20**. *A* **turbine** *is a shaft with a set of blades that spins when a stream of pressurized fluid strikes the blades.* A turbine transfers the mechanical energy of a stream of water, steam, or air to the generator.

Modern turbines that generate electric power are very efficient. They waste little of the mechanical energy used to rotate them.

Reading Check How is a turbine used to produce an electric current?

Transformers—Changing Voltage

Recall that voltage is a measure of the energy transformed by a circuit. A high-voltage circuit transforms more energy than a low-voltage circuit with the same electric current.

Household electrical outlets provide an electric current at 120 V—great enough to cause a dangerous electric shock. However, in large transmission wires, voltage can exceed 500,000 V! At electric power plants, transformers raise the voltage for long-distance transmission, Then, other transformers lower the voltage for household use. *A **transformer** is a device that changes the voltage of an alternating current.*

How a Transformer Works

A transformer consists of two wire coils wrapped around a single iron core, as shown in **Figure 21.** Alternating current in the primary coil produces a continually reversing magnetic field around the core. This changing magnetic field produces alternating current in the secondary coil. If the input voltage of the primary coil increases, the output voltage of the secondary coil increases. And, if the input voltage of the primary coil decreases, the output voltage of the secondary coil decreases.

Step-Down Transformer The output voltage from a transformer also depends on the number of loops in the transformer's coils. If there are fewer loops in the secondary coil than in the primary coil, the transformer's output voltage is less than the input voltage. This type of transformer is a step-down transformer.

Step-Up Transformer If the number of loops in the secondary coil is greater than the number of loops in the primary coil, the output voltage is greater than the input voltage. This type of transformer is called a step-up transformer.

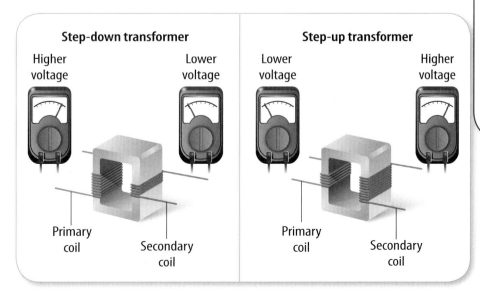

Math Skills

Solve an Equation
You can use math to find the output voltage in a transformer.

Variable	Notation
Number of Loops on Primary Coil	N_p
Number of Loops on Secondary Coil	N_s
Input Voltage to Primary Coil	V_p
Output Voltage from Secondary Coil	V_s

A transformer's primary coil has **200 loops**. Its secondary coil has **3,000 loops**. If its input voltage is **90 V**, what is its output voltage?

$$V_s = V_p \times (N_s \div N_p)$$

Replace the terms in the equation and solve.

$$V_s = 90.0 \text{ V} \times (3{,}000 \text{ loops} / 200 \text{ loops}) = 1{,}350 \text{ V}$$

Practice
A transformer's primary coil has 50 loops. Its secondary coil has 1,500 loops. If its input voltage is 120 V, what is its output voltage?

Review
- Math Practice
- Personal Tutor

Figure 21 A transformer's output voltage can be more or less than its input voltage.

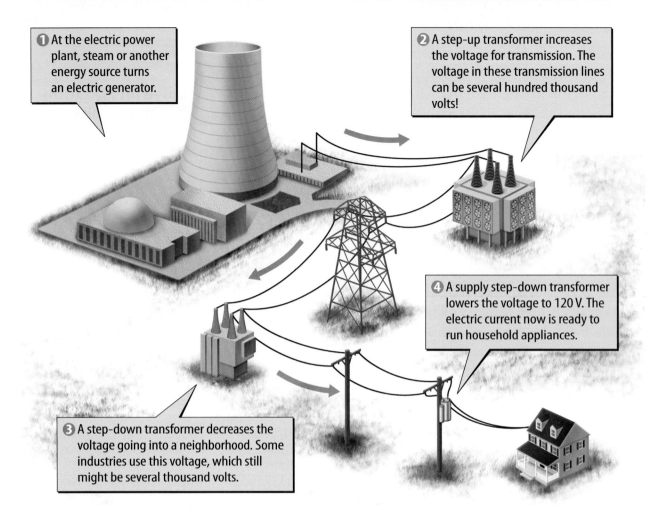

① At the electric power plant, steam or another energy source turns an electric generator.

② A step-up transformer increases the voltage for transmission. The voltage in these transmission lines can be several hundred thousand volts!

③ A step-down transformer decreases the voltage going into a neighborhood. Some industries use this voltage, which still might be several thousand volts.

④ A supply step-down transformer lowers the voltage to 120 V. The electric current now is ready to run household appliances.

Figure 22 Power plants move electric energy through long-distance wires. Transformers raise and lower the voltage of the current as needed by homes and businesses.

Visual Check What type of transformer is used to make electric current safe to enter a home?

Animation

Electric Energy—From a Power Plant to Your Home

Electric energy is transmitted from an electric power plant to homes, as **Figure 22** shows. Transformers control the voltage of the current as it travels from the power plant to homes and businesses. Step-up transformers increase the voltage of the alternating current produced at electric power plants.

High-voltage current is transmitted through transmission wires, or lines. The electric resistance of the wires causes some electric energy to transform to thermal energy. High voltage current in transmission wires reduces the amount of electric energy released as thermal energy to the environment.

However, the high voltage of the current must be reduced for safe use in homes. At neighborhood substations and on utility poles close to your house, step-down transformers decrease the voltage of the current to 120 V.

Key Concept Check How are transformers used to bring an electric current to your home?

Chapter 20
EXPLAIN

Lesson 3 Review

Visual Summary

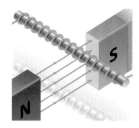
An electric current is produced in a wire coil that passes through a magnetic field.

Electric generators transform mechanical energy to electric energy.

A magnetic field can produce an electric current, and an electric current produces a magnetic field.

FOLDABLES
Use your lesson Foldable to review the lesson. Save your Foldable for the project at the end of the chapter.

What do you think NOW?

You first read the statements below at the beginning of the chapter.

5. A battery produces an electric current that reverses direction in a regular pattern.

6. An electric generator transforms thermal energy into electric energy.

Did you change your mind about whether you agree or disagree with the statements? Rewrite any false statements to make them true.

Use Vocabulary

1. A device that transforms mechanical energy into electric energy is a(n) _____.

2. Electric current that reverses direction in a regular pattern is _____.

3. A device that changes the voltage of an alternating current is a(n) _____.

Understand Key Concepts

4. **Summarize** how a wire and a magnet can be used to produce an electric current.

5. What do step-up transformers do?
   A. control resistance
   B. decrease voltage
   C. increase current
   D. increase voltage

6. **Explain** how an electric generator produces an electric current.

Interpret Graphics

7. **Sequence Events** Copy the graphic organizer below. Show how transformers are used as electric energy is transmitted from a power plant to a home.

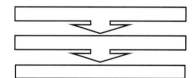

Critical Thinking

8. **Assess** Describe the electric current flowing in the primary coil of a transformer if there is no current flowing in the secondary coil. Explain your answer.

Math Skills

Math Practice

9. A transformer's primary coil has 10,000 loops. The secondary coil has 100 loops. The input voltage is 120 V. What is the output voltage?

3 class periods

Design a Wind-Powered Generator

Materials

nails

strong magnet

cardboard

compass

alligator clip wires

LED lights

Also needed: wood block, 1N34A diode, hammer, metal hangers, enamel-coated wire, drinking straws, small lightbulb (1.5 V), glue or tape, sandpaper

Safety

A generator converts mechanical energy into electric energy. The source of the mechanical energy can be anything from water rushing through a hydroelectric dam to steam produced in a coal-fired or nuclear power plant to wind sweeping across a prairie.

Question

What do you need to design and build a wind-powered generator?

Procedure

1. Read and complete a lab safety form.

2. Build an ammeter: Fold cardboard around a compass to make a base. Wrap 50 loops of wire around the compass. Leave 10-cm tails exposed at each end. Twist the wires to prevent unwinding. Use sandpaper to remove enamel coating from the ends of the wire.

3. Make an electromagnet: Drive a nail 2 cm into a block of wood.

4. Wrap 500 loops of wire around the nail. The coil should be within 1 cm of the head of the nail. Leave 20-cm tails exposed at each end of the coil. Twist the wires to prevent unwinding. Use sandpaper to remove the enamel coating from the ends of the wire.

5. Using alligator clip wires, connect the electromagnet to your ammeter. Position the ammeter at least 1/2 m from the coil. Use additional alligator clip wires if needed.

6. Rotate the ammeter until the needle of the compass is directly under and hidden by the wire wrapping.

7 Glue or tape the magnet to the head of another nail. Position the magnet on the nail so that its north-south axis is perpendicular to the nail.

8 Hold the magnet closely above the electromagnet. Twist the nail with the magnet between your thumb and forefinger so the magnet spins closely over the electromagnet. Do not allow the magnet to touch the nail.

9 Have a lab partner observe the ammeter. Record your team's observations in your Science Journal.

10 Design a structure that captures the wind and will support the nail and the magnet as they spin. Draw your plans in your Science Journal.

11 Have your teacher approve your plans and procedure.

12 Build and test your wind-powered generator.

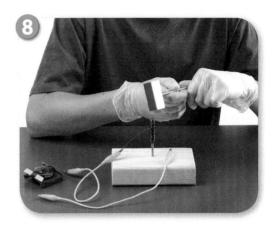

Lab Tips

☑ The turbine creates AC (alternating current) while the ammeter detects DC (direct current). Watch the ammeter, and carefully observe the needle swinging back and forth.

☑ The faster the magnet spins, the more current you will generate, but the harder it will be to measure it using your ammeter. Connect a 1N34A diode in series with the ammeter to convert AC from the generator into DC for the ammeter.

Analyze and Conclude

13 Critique your wind-powered generator. What are the strengths and weaknesses of your generator? How could you revise your design to increase the current produced?

14 Create a flow chart showing the transfer of energy in this system.

15 Explain how your ammeter works.

16 The Big Idea How are moving electric charges and magnetic fields related?

Communicate Your Results

Create a storyboard. Explain the process of making a wind-powered generator and describe how it generates an electric current.

Redesign and improve your wind-powered generator so that it will light a small lightbulb.

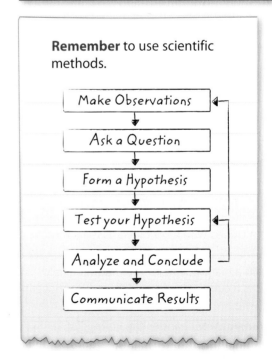

Remember to use scientific methods.

- Make Observations
- Ask a Question
- Form a Hypothesis
- Test your Hypothesis
- Analyze and Conclude
- Communicate Results

Chapter 20 Study Guide

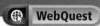

A moving electric charge is surrounded by a magnetic field, and a magnetic field can make electric charges move.

Key Concepts Summary

Lesson 1: Magnets and Magnetic Fields

- A **magnet** is surrounded by a magnetic field that exerts forces on other magnets.
- **Magnetic materials** have **magnetic domains** that point in the same direction.
- The domains of a **temporary magnet** do not remain aligned when removed from a magnetic field. The domains of a **permanent magnet** remain in alignment, even when they are not near another magnet.

Lesson 2: Making Magnets with an Electric Current

- A magnetic field exerts a force on any moving, electrically charged particle, including the charged particles in an electric current.
- The magnetic field around an **electromagnet** can be turned on and off, can reverse direction, and can be made stronger or weaker.
- An **electric motor** contains a permanent magnet and an electromagnet. When an electric current flows in the electromagnet, the forces between the two magnets cause the electromagnet to rotate.

Lesson 3: Making an Electric Current with Magnets

- An electric current is produced when a magnet and a closed wire loop move past each other.
- An **electric generator** has wire loops that rotate within a magnetic field. The magnetic field causes a current of electric charges to move in the wire.
- A **transformer** changes the voltage of an **alternating current.** Step-up transformers raise voltage for cross-country transmission. Step-down transformers lower voltage for household use.

Vocabulary

magnet p. 717
magnetic pole p. 718
magnetic force p. 718
magnetic material p. 721
ferromagnetic element p. 721
magnetic domain p. 722
temporary magnet p. 723
permanent magnet p. 723

electromagnet p. 728
electric motor p. 730

electric generator p. 736
direct current p. 737
alternating current p. 737
turbine p. 738
transformer p. 739

Study Guide

- Personal Tutor
- Vocabulary eGames
- Vocabulary eFlashcards

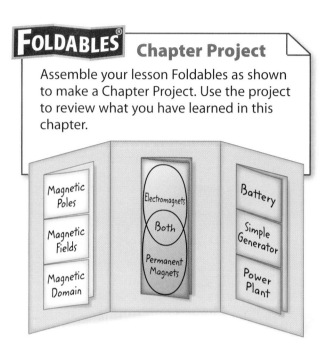

Chapter Project

Assemble your lesson Foldables as shown to make a Chapter Project. Use the project to review what you have learned in this chapter.

Use Vocabulary

1. A(n) _____ can change the mechanical energy of a turbine into electric energy.

2. Similar poles of the _____ of a magnet point in the same direction.

3. An electric generator operates like a(n) _____ in reverse.

4. The magnetic field around a wire reverses direction in a regular pattern if a(n) _____ flows in the wire.

5. A(n) _____ remains a magnet for a long period of time.

6. A core of a ferromagnetic material placed inside the coil of a wire is a(n) _____.

Link Vocabulary and Key Concepts

 Interactive Concept Map

Copy this concept map, and then use vocabulary terms from the previous page and other terms from the chapter to complete the concept map.

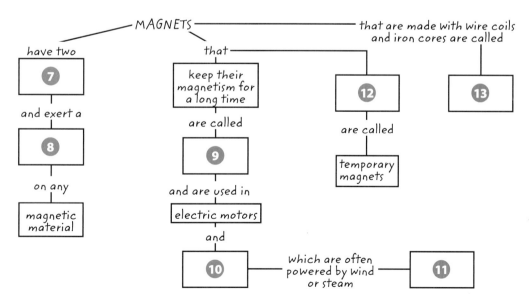

Chapter 20 Study Guide • 745

Chapter 20 Review

Understand Key Concepts

1. When the current in an electromagnet increases, what does its magnetic field do?
    A. changes direction
    B. does not change
    C. gets stronger
    D. gets weaker

2. What is the function of the commutator in an electric motor?
    A. It decreases the current in the coil.
    B. It reverses the current in the coil.
    C. It reverses the poles of the permanent magnet.
    D. It weakens the field of the permanent magnet.

3. When a bar magnet moves in a wire loop, what does the wire loop produce?
    A. an electric current
    B. electrical resistance
    C. ferromagnetic elements
    D. magnetic domains

4. Which is NOT a ferromagnetic element?
    A. aluminum
    B. cobalt
    C. iron
    D. nickel

5. Which best describes the force between the two magnets shown below?

    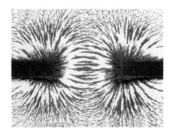

    A. alternating
    B. attractive
    C. direct
    D. repulsive

6. Where on a magnet is the magnetic field the strongest?
    A. both magnetic poles
    B. center of the magnet
    C. magnetic north pole
    D. magnetic south pole

7. In the figure below, how does the magnetic force on the wire change if the current in the wire changes direction?

    A. reverses
    B. strengthens
    C. weakens
    D. remains constant

8. Which energy transformation occurs in a generator?
    A. electric energy to light energy
    B. electric energy to mechanical energy
    C. mechanical energy to electric energy
    D. thermal energy to electric energy

9. Which describes the direction of the magnetic field lines around a magnet?
    A. start at center, end at each pole
    B. start at each pole, end at center
    C. start at north pole, end at south pole
    D. start at south pole, end at north pole

10. Which best describes a nonmagnetic material?
    A. The magnetic poles of its atoms point in different directions.
    B. The magnetic poles of its atoms point in the same direction.
    C. The magnetic poles of its magnetic domains point in different directions.
    D. The magnetic poles of its magnetic domains point in the same direction.

Chapter Review

Assessment
Online Test Practice

Critical Thinking

11 **Predict** A step-up transformer is used to connect a lightbulb to a wall outlet. What happens to the brightness of the lightbulb if the transformer is replaced with a step-down transformer?

12 **Defend** One electromagnet has a wood core. Another electromagnet has an iron core. Which has the stronger magnetic field? Explain your answer.

13 **Assess** Compare and contrast the ways magnetic poles interact with the ways electric charges interact.

14 **Suggest** How could you determine a car door is made of a nonmagnetic material, such as plastic?

15 **Suggest** What would you do to a wire to make it attract to a magnet?

16 **Evaluate** The figure below shows the field lines of a magnet. Where is the magnet's north pole? Explain your answer.

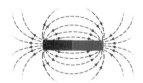

17 **Recommend** How could you increase the number of times per second that an alternating current from a generator changes direction?

18 **Hypothesize** An alternating current flows in an electromagnet. If the electromagnet is placed in a wire coil, will there be an electric current in the wire coil? Explain your answer.

Writing in Science

19 **Write** a short essay describing the different devices you use in a typical day that have electric motors. Include how the electric motors are used to make something move.

REVIEW THE BIG IDEA

20 How can a magnetic field cause a current to flow in a wire coil?

21 How are electric charges and magnetic fields related?

Math Skills

Review Math Practice

Solve an Equation

22 A transformer has 300 loops on its primary coil and 90,000 loops on its secondary coil. The input voltage is 60 V. What is the output voltage?

23 A transformer has 80 loops on its primary coil and 1,200 loops on its secondary coil. The input voltage is 60 V. What is the output voltage?

24 The primary coil of a transformer has 150 loops and has a 120 V primary source. How many loops should the secondary coil have to produce an output voltage of 600 V?

Standardized Test Practice

Record your answers on the answer sheet provided by your teacher or on a sheet of paper.

Multiple Choice

1 Which statement about magnets is true?
   A Two north poles attract each other.
   B Two south poles attract each other.
   C A north pole and a south pole attract each other.
   D A north pole and a south pole repel each other.

Use the images below to answer question 2.

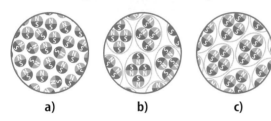

a)   b)   c)

2 Which statement best describes these models?
   A Only A is a magnet.
   B Only C is a magnet.
   C A and B are magnets.
   D A and C are magnets.

3 Why does a compass needle move if held near a current-carrying wire?
   A A current-carrying wire will cause any object to move.
   B A compass needle moves when it is near any piece of metal wire.
   C Both the compass needle and the current-carrying wire have magnetic fields.
   D The compass needle and the current-carrying wire have the same electric charge.

Use the image below to answer question 4.

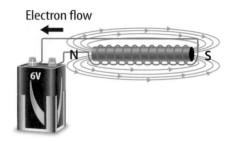

4 Which describes what will happen if the iron core is removed from the electromagnet?
   A It is no longer a magnet.
   B It is a stronger magnet.
   C It is a weaker magnet.
   D It loses its north and south poles.

5 What causes the poles of the electromagnet in an electric motor to reverse?
   A a change in the direction of the current
   B a change in the strength of the magnetic field
   C a change in the shape of the current-carrying wire
   D a change in the position of the permanent magnet

6 What happens when a magnet passes through a closed loop of wire?
   A An electric current is generated in the wire.
   B Any nearby magnets gain an electric charge.
   C The looped part of the wire no longer is attracted to magnets.
   D Any device in the circuit becomes a permanent magnet.

Standardized Test Practice

7 How do an electric motor and an electric generator differ?
   A One uses a permanent magnet, while the other uses temporary magnets.
   B One produces electric energy, while the other produces thermal energy.
   C One contains moving parts, while the other has only parts that do not move.
   D One uses an electric current to produce motion, while the other uses motion to produce an electric current.

Use the figure below to answer questions 8 and 9.

8 What would be the use for this device?
   A to generate electric energy
   B to increase the voltage
   C to make an electric motor
   D to reduce the voltage

9 For the device, which statement is true?
   A The primary voltage is greater than the secondary voltage.
   B The primary voltage is less than the secondary voltage.
   C The primary voltage is the same secondary voltage.
   D The voltages alternate high to low.

Constructed Response

10 When you touch a magnet to a steel nail, the nail attracts steel paper clips. Explain why this does not happen when you use an aluminum nail.

Use the figure below to answer questions 11 and 12.

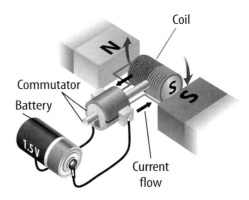

11 Using the terms in the diagram, identify what this device is and describe what it does.

12 What would happen to the device if the wire coil were removed?

13 How are transformers used as an electric current travels between a power plant and a home?

14 Compare and contrast the components and functions of an electric motor and an electric generator.

NEED EXTRA HELP?														
If You Missed Question...	1	2	3	4	5	6	7	8	9	10	11	12	13	14
Go to Lesson...	1	1	2	2	2	3	3	3	3	1	2	2	3	3

Student Resources

For Students and Parents/Guardians

These resources are designed to help you achieve success in science. You will find useful information on laboratory safety, math skills, and science skills. In addition, science reference materials are found in the Reference Handbook. You'll find the information you need to learn and sharpen your skills in these resources.

Table of Contents

Science Skill Handbook SR-2

Scientific Methods SR-2
   Identify a Question.. SR-2
   Gather and Organize Information.................................... SR-2
   Form a Hypothesis... SR-5
   Test the Hypothesis SR-6
   Collect Data SR-6
   Analyze the Data.. SR-9
   Draw Conclusions SR-10
   Communicate... SR-10

Safety Symbols SR-11
Safety in the Science Laboratory.............................. SR-12
   General Safety Rules SR-12
   Prevent Accidents.. SR-12
   Laboratory Work SR-13
   Emergencies.. SR-13

Math Skill Handbook SR-14

Math Review.. SR-14
   Use Fractions SR-14
   Use Ratios SR-17
   Use Decimals SR-17
   Use Proportions... SR-18
   Use Percentages SR-19
   Solve One-Step Equations.. SR-19
   Use Statistics.. SR-20
   Use Geometry SR-21

Science Application SR-24
   Measure in SI SR-24
   Dimensional Analysis.. SR-24
   Precision and Significant Digits SR-26
   Scientific Notation SR-26
   Make and Use Graphs SR-27

Foldables Handbook SR-29

Reference Handbook SR-40
   Periodic Table of the Elements...................................... SR-40

Glossary G-2

Index I-2

Credits C-2

Science Skill Handbook

Scientific Methods

Scientists use an orderly approach called the scientific method to solve problems. This includes organizing and recording data so others can understand them. Scientists use many variations in this method when they solve problems.

Identify a Question

The first step in a scientific investigation or experiment is to identify a question to be answered or a problem to be solved. For example, you might ask which gasoline is the most efficient.

Gather and Organize Information

After you have identified your question, begin gathering and organizing information. There are many ways to gather information, such as researching in a library, interviewing those knowledgeable about the subject, and testing and working in the laboratory and field. Fieldwork is investigations and observations done outside of a laboratory.

Researching Information Before moving in a new direction, it is important to gather the information that already is known about the subject. Start by asking yourself questions to determine exactly what you need to know. Then you will look for the information in various reference sources, like the student is doing in **Figure 1.** Some sources may include textbooks, encyclopedias, government documents, professional journals, science magazines, and the Internet. Always list the sources of your information.

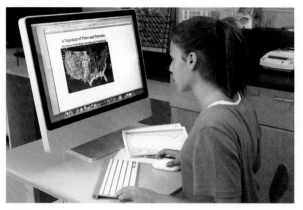

Figure 1 The Internet can be a valuable research tool.

Evaluate Sources of Information Not all sources of information are reliable. You should evaluate all of your sources of information, and use only those you know to be dependable. For example, if you are researching ways to make homes more energy efficient, a site written by the U.S. Department of Energy would be more reliable than a site written by a company that is trying to sell a new type of weatherproofing material. Also, remember that research always is changing. Consult the most current resources available to you. For example, a 1985 resource about saving energy would not reflect the most recent findings.

Sometimes scientists use data that they did not collect themselves, or conclusions drawn by other researchers. This data must be evaluated carefully. Ask questions about how the data were obtained, if the investigation was carried out properly, and if it has been duplicated exactly with the same results. Would you reach the same conclusion from the data? Only when you have confidence in the data can you believe it is true and feel comfortable using it.

Interpret Scientific Illustrations As you research a topic in science, you will see drawings, diagrams, and photographs to help you understand what you read. Some illustrations are included to help you understand an idea that you can't see easily by yourself, like the tiny particles in an atom in **Figure 2**. A drawing helps many people to remember details more easily and provides examples that clarify difficult concepts or give additional information about the topic you are studying. Most illustrations have labels or a caption to identify or to provide more information.

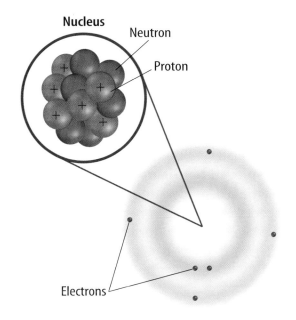

Figure 2 This drawing shows an atom of carbon with its six protons, six neutrons, and six electrons.

Concept Maps One way to organize data is to draw a diagram that shows relationships among ideas (or concepts). A concept map can help make the meanings of ideas and terms more clear, and help you understand and remember what you are studying. Concept maps are useful for breaking large concepts down into smaller parts, making learning easier.

Network Tree A type of concept map that not only shows a relationship, but how the concepts are related is a network tree, shown in **Figure 3**. In a network tree, the words are written in the ovals, while the description of the type of relationship is written across the connecting lines.

When constructing a network tree, write down the topic and all major topics on separate pieces of paper or notecards. Then arrange them in order from general to specific. Branch the related concepts from the major concept and describe the relationship on the connecting line. Continue to more specific concepts until finished.

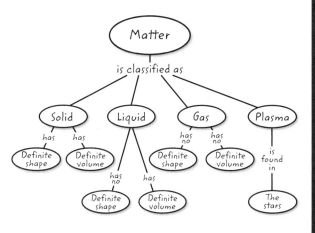

Figure 3 A network tree shows how concepts or objects are related.

Events Chain Another type of concept map is an events chain. Sometimes called a flow chart, it models the order or sequence of items. An events chain can be used to describe a sequence of events, the steps in a procedure, or the stages of a process.

When making an events chain, first find the one event that starts the chain. This event is called the initiating event. Then, find the next event and continue until the outcome is reached, as shown in **Figure 4** on the next page.

Science Skill Handbook • **SR-3**

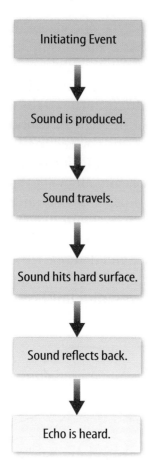

Figure 4 Events-chain concept maps show the order of steps in a process or event. This concept map shows how a sound makes an echo.

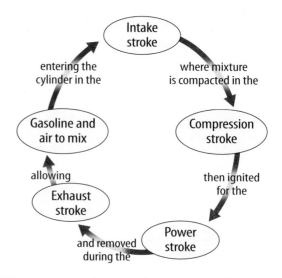

Figure 5 A cycle map shows events that occur in a cycle.

Cycle Map A specific type of events chain is a cycle map. It is used when the series of events do not produce a final outcome, but instead relate back to the beginning event, such as in **Figure 5**. Therefore, the cycle repeats itself.

To make a cycle map, first decide what event is the beginning event. This is also called the initiating event. Then list the next events in the order that they occur, with the last event relating back to the initiating event. Words can be written between the events that describe what happens from one event to the next. The number of events in a cycle map can vary, but usually contain three or more events.

Spider Map A type of concept map that you can use for brainstorming is the spider map. When you have a central idea, you might find that you have a jumble of ideas that relate to it but are not necessarily clearly related to each other. The spider map on sound in **Figure 6** shows that if you write these ideas outside the main concept, then you can begin to separate and group unrelated terms so they become more useful.

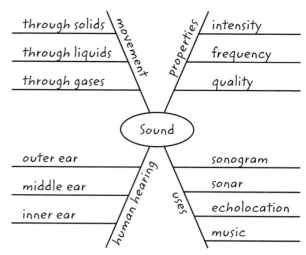

Figure 6 A spider map allows you to list ideas that relate to a central topic but not necessarily to one another.

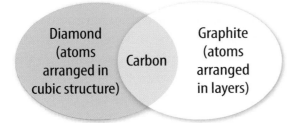

Figure 7 This Venn diagram compares and contrasts two substances made from carbon.

Venn Diagram To illustrate how two subjects compare and contrast you can use a Venn diagram. You can see the characteristics that the subjects have in common and those that they do not, shown in **Figure 7**.

To create a Venn diagram, draw two overlapping ovals that are big enough to write in. List the characteristics unique to one subject in one oval, and the characteristics of the other subject in the other oval. The characteristics in common are listed in the overlapping section.

Make and Use Tables One way to organize information so it is easier to understand is to use a table. Tables can contain numbers, words, or both.

To make a table, list the items to be compared in the first column and the characteristics to be compared in the first row. The title should clearly indicate the content of the table, and the column or row heads should be clear. Notice that in **Table 1** the units are included.

Table 1 Recyclables Collected During Week			
Day of Week	Paper (kg)	Aluminum (kg)	Glass (kg)
Monday	5.0	4.0	12.0
Wednesday	4.0	1.0	10.0
Friday	2.5	2.0	10.0

Make a Model One way to help you better understand the parts of a structure, the way a process works, or to show things too large or small for viewing is to make a model. For example, an atomic model made of a plastic-ball nucleus and chenille stem electron shells can help you visualize how the parts of an atom relate to each other. Other types of models can be devised on a computer or represented by equations.

Form a Hypothesis

A possible explanation based on previous knowledge and observations is called a hypothesis. After researching gasoline types and recalling previous experiences in your family's car you form a hypothesis—our car runs more efficiently because we use premium gasoline. To be valid, a hypothesis has to be something you can test by using an investigation.

Predict When you apply a hypothesis to a specific situation, you predict something about that situation. A prediction makes a statement in advance, based on prior observation, experience, or scientific reasoning. People use predictions to make everyday decisions. Scientists test predictions by performing investigations. Based on previous observations and experiences, you might form a prediction that cars are more efficient with premium gasoline. The prediction can be tested in an investigation.

Design an Experiment A scientist needs to make many decisions before beginning an investigation. Some of these include: how to carry out the investigation, what steps to follow, how to record the data, and how the investigation will answer the question. It also is important to address any safety concerns.

Test the Hypothesis

Now that you have formed your hypothesis, you need to test it. Using an investigation, you will make observations and collect data, or information. This data might either support or not support your hypothesis. Scientists collect and organize data as numbers and descriptions.

Follow a Procedure In order to know what materials to use, as well as how and in what order to use them, you must follow a procedure. **Figure 8** shows a procedure you might follow to test your hypothesis.

Procedure

Step 1 Use regular gasoline for two weeks.

Step 2 Record the number of kilometers between fill-ups and the amount of gasoline used.

Step 3 Switch to premium gasoline for two weeks.

Step 4 Record the number of kilometers between fill-ups and the amount of gasoline used.

Figure 8 A procedure tells you what to do step-by-step.

Identify and Manipulate Variables and Controls In any experiment, it is important to keep everything the same except for the item you are testing. The one factor you change is called the independent variable. The change that results is the dependent variable. Make sure you have only one independent variable, to assure yourself of the cause of the changes you observe in the dependent variable. For example, in your gasoline experiment the type of fuel is the independent variable. The dependent variable is the efficiency.

Many experiments also have a control—an individual instance or experimental subject for which the independent variable is not changed. You can then compare the test results to the control results. To design a control you can have two cars of the same type. The control car uses regular gasoline for four weeks. After you are done with the test, you can compare the experimental results to the control results.

Collect Data

Whether you are carrying out an investigation or a short observational experiment, you will collect data, as shown in **Figure 9**. Scientists collect data as numbers and descriptions and organize them in specific ways.

Observe Scientists observe items and events, then record what they see. When they use only words to describe an observation, it is called qualitative data. Scientists' observations also can describe how much there is of something. These observations use numbers, as well as words, in the description and are called quantitative data. For example, if a sample of the element gold is described as being "shiny and very dense" the data are qualitative. Quantitative data on this sample of gold might include "a mass of 30 g and a density of 19.3 g/cm^3."

Figure 9 Collecting data is one way to gather information directly.

Figure 10 Record data neatly and clearly so it is easy to understand.

When you make observations you should examine the entire object or situation first, and then look carefully for details. It is important to record observations accurately and completely. Always record your notes immediately as you make them, so you do not miss details or make a mistake when recording results from memory. Never put unidentified observations on scraps of paper. Instead they should be recorded in a notebook, like the one in **Figure 10.** Write your data neatly so you can easily read it later. At each point in the experiment, record your observations and label them. That way, you will not have to determine what the figures mean when you look at your notes later. Set up any tables that you will need to use ahead of time, so you can record any observations right away. Remember to avoid bias when collecting data by not including personal thoughts when you record observations. Record only what you observe.

Estimate Scientific work also involves estimating. To estimate is to make a judgment about the size or the number of something without measuring or counting. This is important when the number or size of an object or population is too large or too difficult to accurately count or measure.

Sample Scientists may use a sample or a portion of the total number as a type of estimation. To sample is to take a small, representative portion of the objects or organisms of a population for research. By making careful observations or manipulating variables within that portion of the group, information is discovered and conclusions are drawn that might apply to the whole population. A poorly chosen sample can be unrepresentative of the whole. If you were trying to determine the rainfall in an area, it would not be best to take a rainfall sample from under a tree.

Measure You use measurements every day. Scientists also take measurements when collecting data. When taking measurements, it is important to know how to use measuring tools properly. Accuracy also is important.

Length To measure length, the distance between two points, scientists use meters. Smaller measurements might be measured in centimeters or millimeters.

Length is measured using a metric ruler or meterstick. When using a metric ruler, line up the 0-cm mark with the end of the object being measured and read the number of the unit where the object ends. Look at the metric ruler shown in **Figure 11.** The centimeter lines are the long, numbered lines, and the shorter lines are millimeter lines. In this instance, the length would be 4.50 cm.

Figure 11 This metric ruler has centimeter and millimeter divisions.

Mass The SI unit for mass is the kilogram (kg). Scientists can measure mass using units formed by adding metric prefixes to the unit gram (g), such as milligram (mg). To measure mass, you might use a triple-beam balance similar to the one shown in **Figure 12.** The balance has a pan on one side and a set of beams on the other side. Each beam has a rider that slides on the beam.

When using a triple-beam balance, place an object on the pan. Slide the largest rider along its beam until the pointer drops below zero. Then move it back one notch. Repeat the process for each rider proceeding from the larger to smaller until the pointer swings an equal distance above and below the zero point. Sum the masses on each beam to find the mass of the object. Move all riders back to zero when finished.

Instead of putting materials directly on the balance, scientists often take a tare of a container. A tare is the mass of a container into which objects or substances are placed for measuring their masses. To find the mass of objects or substances, find the mass of a clean container. Remove the container from the pan, and place the object or substances in the container. Find the mass of the container with the materials in it. Subtract the mass of the empty container from the mass of the filled container to find the mass of the materials you are using.

Figure 12 A triple-beam balance is used to determine the mass of an object.

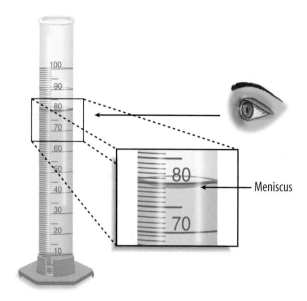

Figure 13 Graduated cylinders measure liquid volume.

Liquid Volume To measure liquids, the unit used is the liter. When a smaller unit is needed, scientists might use a milliliter. Because a milliliter takes up the volume of a cube measuring 1 cm on each side it also can be called a cubic centimeter ($cm^3 = cm \times cm \times cm$).

You can use beakers and graduated cylinders to measure liquid volume. A graduated cylinder, shown in **Figure 13,** is marked from bottom to top in milliliters. In lab, you might use a 10-mL graduated cylinder or a 100-mL graduated cylinder. When measuring liquids, notice that the liquid has a curved surface. Look at the surface at eye level, and measure the bottom of the curve. This is called the meniscus. The graduated cylinder in **Figure 13** contains 79.0 mL, or 79.0 cm^3, of a liquid.

Temperature Scientists often measure temperature using the Celsius scale. Pure water has a freezing point of 0°C and boiling point of 100°C. The unit of measurement is degrees Celsius. Two other scales often used are the Fahrenheit and Kelvin scales.

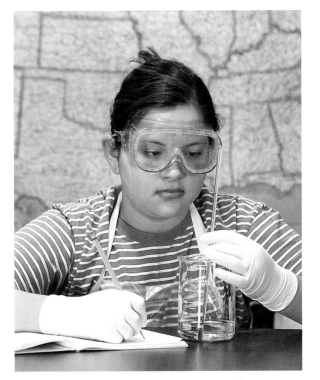

Figure 14 A thermometer measures the temperature of an object.

Scientists use a thermometer to measure temperature. Most thermometers in a laboratory are glass tubes with a bulb at the bottom end containing a liquid such as colored alcohol. The liquid rises or falls with a change in temperature. To read a glass thermometer like the thermometer in **Figure 14,** rotate it slowly until a red line appears. Read the temperature where the red line ends.

Form Operational Definitions An operational definition defines an object by how it functions, works, or behaves. For example, when you are playing hide and seek and a tree is home base, you have created an operational definition for a tree.

Objects can have more than one operational definition. For example, a ruler can be defined as a tool that measures the length of an object (how it is used). It can also be a tool with a series of marks used as a standard when measuring (how it works).

Analyze the Data

To determine the meaning of your observations and investigation results, you will need to look for patterns in the data. Then you must think critically to determine what the data mean. Scientists use several approaches when they analyze the data they have collected and recorded. Each approach is useful for identifying specific patterns.

Interpret Data The word *interpret* means "to explain the meaning of something." When analyzing data from an experiment, try to find out what the data show. Identify the control group and the test group to see whether changes in the independent variable have had an effect. Look for differences in the dependent variable between the control and test groups.

Classify Sorting objects or events into groups based on common features is called classifying. When classifying, first observe the objects or events to be classified. Then select one feature that is shared by some members in the group, but not by all. Place those members that share that feature in a subgroup. You can classify members into smaller and smaller subgroups based on characteristics. Remember that when you classify, you are grouping objects or events for a purpose. Keep your purpose in mind as you select the features to form groups and subgroups.

Compare and Contrast Observations can be analyzed by noting the similarities and differences between two or more objects or events that you observe. When you look at objects or events to see how they are similar, you are comparing them. Contrasting is looking for differences in objects or events.

Science Skill Handbook • **SR-9**

Recognize Cause and Effect A cause is a reason for an action or condition. The effect is that action or condition. When two events happen together, it is not necessarily true that one event caused the other. Scientists must design a controlled investigation to recognize the exact cause and effect.

Draw Conclusions

When scientists have analyzed the data they collected, they proceed to draw conclusions about the data. These conclusions are sometimes stated in words similar to the hypothesis that you formed earlier. They may confirm a hypothesis, or lead you to a new hypothesis.

Infer Scientists often make inferences based on their observations. An inference is an attempt to explain observations or to indicate a cause. An inference is not a fact, but a logical conclusion that needs further investigation. For example, you may infer that a fire has caused smoke. Until you investigate, however, you do not know for sure.

Apply When you draw a conclusion, you must apply those conclusions to determine whether the data supports the hypothesis. If your data do not support your hypothesis, it does not mean that the hypothesis is wrong. It means only that the result of the investigation did not support the hypothesis. Maybe the experiment needs to be redesigned, or some of the initial observations on which the hypothesis was based were incomplete or biased. Perhaps more observation or research is needed to refine your hypothesis. A successful investigation does not always come out the way you originally predicted.

Avoid Bias Sometimes a scientific investigation involves making judgments. When you make a judgment, you form an opinion. It is important to be honest and not to allow any expectations of results to bias your judgments. This is important throughout the entire investigation, from researching to collecting data to drawing conclusions.

Communicate

The communication of ideas is an important part of the work of scientists. A discovery that is not reported will not advance the scientific community's understanding or knowledge. Communication among scientists also is important as a way of improving their investigations.

Scientists communicate in many ways, from writing articles in journals and magazines that explain their investigations and experiments, to announcing important discoveries on television and radio. Scientists also share ideas with colleagues on the Internet or present them as lectures, like the student is doing in **Figure 15**.

Figure 15 A student communicates to his peers about his investigation.

These safety symbols are used in laboratory and field investigations in this book to indicate possible hazards. Learn the meaning of each symbol and refer to this page often. *Remember to wash your hands thoroughly after completing lab procedures.*

PROTECTIVE EQUIPMENT Do not begin any lab without the proper protection equipment.

 GOGGLES Proper eye protection must be worn when performing or observing science activities that involve items or conditions as listed below.

 APRON Wear an approved apron when using substances that could stain, wet, or destroy cloth.

 SOAP Wash hands with soap and water before removing goggles and after all lab activities.

 GLOVES Wear gloves when working with biological materials, chemicals, animals, or materials that can stain or irritate hands.

LABORATORY HAZARDS

Symbols	Potential Hazards	Precaution	Response
DISPOSAL	contamination of classroom or environment due to improper disposal of materials such as chemicals and live specimens	• DO NOT dispose of hazardous materials in the sink or trash can. • Dispose of wastes as directed by your teacher.	• If hazardous materials are disposed of improperly, notify your teacher immediately.
EXTREME TEMPERATURE	skin burns due to extremely hot or cold materials such as hot glass, liquids, or metals; liquid nitrogen; dry ice	• Use proper protective equipment, such as hot mitts and/or tongs, when handling objects with extreme temperatures.	• If injury occurs, notify your teacher immediately.
SHARP OBJECTS	punctures or cuts from sharp objects such as razor blades, pins, scalpels, and broken glass	• Handle glassware carefully to avoid breakage. • Walk with sharp objects pointed downward, away from you and others.	• If broken glass or injury occurs, notify your teacher immediately.
ELECTRICAL	electric shock or skin burn due to improper grounding, short circuits, liquid spills, or exposed wires	• Check condition of wires and apparatus for fraying or uninsulated wires, and broken or cracked equipment. • Use only GFCI-protected outlets	• DO NOT attempt to fix electrical problems. Notify your teacher immediately.
CHEMICAL	skin irritation or burns, breathing difficulty, and/or poisoning due to touching, swallowing, or inhalation of chemicals such as acids, bases, bleach, metal compounds, iodine, poinsettias, pollen, ammonia, acetone, nail polish remover, heated chemicals, mothballs, and any other chemicals labeled or known to be dangerous	• Wear proper protective equipment such as goggles, apron, and gloves when using chemicals. • Ensure proper room ventilation or use a fume hood when using materials that produce fumes. • NEVER smell fumes directly. • NEVER taste or eat any material in the laboratory.	• If contact occurs, immediately flush affected area with water and notify your teacher. • If a spill occurs, leave the area immediately and notify your teacher.
FLAMMABLE	unexpected fire due to liquids or gases that ignite easily such as rubbing alcohol	• Avoid open flames, sparks, or heat when flammable liquids are present.	• If a fire occurs, leave the area immediately and notify your teacher.
OPEN FLAME	burns or fire due to open flame from matches, Bunsen burners, or burning materials	• Tie back loose hair and clothing. • Keep flame away from all materials. • Follow teacher instructions when lighting and extinguishing flames. • Use proper protection, such as hot mitts or tongs, when handling hot objects.	• If a fire occurs, leave the area immediately and notify your teacher.
ANIMAL SAFETY	injury to or from laboratory animals	• Wear proper protective equipment such as gloves, apron, and goggles when working with animals. • Wash hands after handling animals.	• If injury occurs, notify your teacher immediately.
BIOLOGICAL	infection or adverse reaction due to contact with organisms such as bacteria, fungi, and biological materials such as blood, animal or plant materials	• Wear proper protective equipment such as gloves, goggles, and apron when working with biological materials. • Avoid skin contact with an organism or any part of the organism. • Wash hands after handling organisms.	• If contact occurs, wash the affected area and notify your teacher immediately.
FUME	breathing difficulties from inhalation of fumes from substances such as ammonia, acetone, nail polish remover, heated chemicals, and mothballs	• Wear goggles, apron, and gloves. • Ensure proper room ventilation or use a fume hood when using substances that produce fumes. • NEVER smell fumes directly.	• If a spill occurs, leave area and notify your teacher immediately.
IRRITANT	irritation of skin, mucous membranes, or respiratory tract due to materials such as acids, bases, bleach, pollen, mothballs, steel wool, and potassium permanganate	• Wear goggles, apron, and gloves. • Wear a dust mask to protect against fine particles.	• If skin contact occurs, immediately flush the affected area with water and notify your teacher.
RADIOACTIVE	excessive exposure from alpha, beta, and gamma particles	• Remove gloves and wash hands with soap and water before removing remainder of protective equipment.	• If cracks or holes are found in the container, notify your teacher immediately.

Safety in the Science Laboratory

Introduction to Science Safety

The science laboratory is a safe place to work if you follow standard safety procedures. Being responsible for your own safety helps to make the entire laboratory a safer place for everyone. When performing any lab, read and apply the caution statements and safety symbol listed at the beginning of the lab.

General Safety Rules

1. Complete the *Lab Safety Form* or other safety contract BEFORE starting any science lab.
2. Study the procedure. Ask your teacher any questions. Be sure you understand safety symbols shown on the page.
3. Notify your teacher about allergies or other health conditions that can affect your participation in a lab.
4. Learn and follow use and safety procedures for your equipment. If unsure, ask your teacher.

5. Never eat, drink, chew gum, apply cosmetics, or do any personal grooming in the lab. Never use lab glassware as food or drink containers. Keep your hands away from your face and mouth.
6. Know the location and proper use of the safety shower, eye wash, fire blanket, and fire alarm.

Prevent Accidents

1. Use the safety equipment provided to you. Goggles and a safety apron should be worn during investigations.
2. Do NOT use hair spray, mousse, or other flammable hair products. Tie back long hair and tie down loose clothing.
3. Do NOT wear sandals or other open-toed shoes in the lab.
4. Remove jewelry on hands and wrists. Loose jewelry, such as chains and long necklaces, should be removed to prevent them from getting caught in equipment.
5. Do not taste any substances or draw any material into a tube with your mouth.
6. Proper behavior is expected in the lab. Practical jokes and fooling around can lead to accidents and injury.
7. Keep your work area uncluttered.

Laboratory Work

1. Collect and carry all equipment and materials to your work area before beginning a lab.
2. Remain in your own work area unless given permission by your teacher to leave it.

3. Always slant test tubes away from yourself and others when heating them, adding substances to them, or rinsing them.
4. If instructed to smell a substance in a container, hold the container a short distance away and fan vapors toward your nose.
5. Do NOT substitute other chemicals/substances for those in the materials list unless instructed to do so by your teacher.
6. Do NOT take any materials or chemicals outside of the laboratory.
7. Stay out of storage areas unless instructed to be there and supervised by your teacher.

Laboratory Cleanup

1. Turn off all burners, water, and gas, and disconnect all electrical devices.
2. Clean all pieces of equipment and return all materials to their proper places.
3. Dispose of chemicals and other materials as directed by your teacher. Place broken glass and solid substances in the proper containers. Never discard materials in the sink.
4. Clean your work area.
5. Wash your hands with soap and water thoroughly BEFORE removing your goggles.

Emergencies

1. Report any fire, electrical shock, glassware breakage, spill, or injury, no matter how small, to your teacher immediately. Follow his or her instructions.
2. If your clothing should catch fire, STOP, DROP, and ROLL. If possible, smother it with the fire blanket or get under a safety shower. NEVER RUN.
3. If a fire should occur, turn off all gas and leave the room according to established procedures.
4. In most instances, your teacher will clean up spills. Do NOT attempt to clean up spills unless you are given permission and instructions to do so.
5. If chemicals come into contact with your eyes or skin, notify your teacher immediately. Use the eyewash, or flush your skin or eyes with large quantities of water.
6. The fire extinguisher and first-aid kit should only be used by your teacher unless it is an extreme emergency and you have been given permission.
7. If someone is injured or becomes ill, only a professional medical provider or someone certified in first aid should perform first-aid procedures.

Science Skill Handbook • **SR-13**

Math Skill Handbook

Math Review

Use Fractions

A fraction compares a part to a whole. In the fraction $\frac{2}{3}$, the 2 represents the part and is the numerator. The 3 represents the whole and is the denominator.

Reduce Fractions To reduce a fraction, you must find the largest factor that is common to both the numerator and the denominator, the greatest common factor (GCF). Divide both numbers by the GCF. The fraction has then been reduced, or it is in its simplest form.

Example

Twelve of the 20 chemicals in the science lab are in powder form. What fraction of the chemicals used in the lab are in powder form?

Step 1 Write the fraction.

$$\frac{part}{whole} = \frac{12}{20}$$

Step 2 To find the GCF of the numerator and denominator, list all of the factors of each number.

Factors of 12: 1, 2, 3, 4, 6, 12 (the numbers that divide evenly into 12)

Factors of 20: 1, 2, 4, 5, 10, 20 (the numbers that divide evenly into 20)

Step 3 List the common factors.

1, 2, 4

Step 4 Choose the greatest factor in the list. The GCF of 12 and 20 is 4.

Step 5 Divide the numerator and denominator by the GCF.

$$\frac{12 \div 4}{20 \div 4} = \frac{3}{5}$$

In the lab, $\frac{3}{5}$ of the chemicals are in powder form.

Practice Problem At an amusement park, 66 of 90 rides have a height restriction. What fraction of the rides, in its simplest form, has a height restriction?

Add and Subtract Fractions with Like Denominators To add or subtract fractions with the same denominator, add or subtract the numerators and write the sum or difference over the denominator. After finding the sum or difference, find the simplest form for your fraction.

Example 1

In the forest outside your house, $\frac{1}{8}$ of the animals are rabbits, $\frac{3}{8}$ are squirrels, and the remainder are birds and insects. How many are mammals?

Step 1 Add the numerators.

$$\frac{1}{8} + \frac{3}{8} = \frac{(1+3)}{8} = \frac{4}{8}$$

Step 2 Find the GCF.

$\frac{4}{8}$ (GCF, 4)

Step 3 Divide the numerator and denominator by the GCF.

$$\frac{4 \div 4}{8 \div 4} = \frac{1}{2}$$

$\frac{1}{2}$ of the animals are mammals.

Example 2

If $\frac{7}{16}$ of the Earth is covered by freshwater, and $\frac{1}{16}$ of that is in glaciers, how much freshwater is not frozen?

Step 1 Subtract the numerators.

$$\frac{7}{16} - \frac{1}{16} = \frac{(7-1)}{16} = \frac{6}{16}$$

Step 2 Find the GCF.

$\frac{6}{16}$ (GCF, 2)

Step 3 Divide the numerator and denominator by the GCF.

$$\frac{6 \div 2}{16 \div 2} = \frac{3}{8}$$

$\frac{3}{8}$ of the freshwater is not frozen.

Practice Problem A bicycle rider is riding at a rate of 15 km/h for $\frac{4}{9}$ of his ride, 10 km/h for $\frac{2}{9}$ of his ride, and 8 km/h for the remainder of the ride. How much of his ride is he riding at a rate greater than 8 km/h?

Add and Subtract Fractions with Unlike Denominators To add or subtract fractions with unlike denominators, first find the least common denominator (LCD). This is the smallest number that is a common multiple of both denominators. Rename each fraction with the LCD, and then add or subtract. Find the simplest form if necessary.

Example 1

A chemist makes a paste that is $\frac{1}{2}$ table salt (NaCl), $\frac{1}{3}$ sugar ($C_6H_{12}O_6$), and the remainder is water (H_2O). How much of the paste is a solid?

Step 1 Find the LCD of the fractions.

$\frac{1}{2} + \frac{1}{3}$ (LCD, 6)

Step 2 Rename each numerator and each denominator with the LCD.

Step 3 Add the numerators.

$\frac{3}{6} + \frac{2}{6} = \frac{(3+2)}{6} = \frac{5}{6}$

$\frac{5}{6}$ of the paste is a solid.

Example 2

The average precipitation in Grand Junction, CO, is $\frac{7}{10}$ inch in November, and $\frac{3}{5}$ inch in December. What is the total average precipitation?

Step 1 Find the LCD of the fractions.

$\frac{7}{10} + \frac{3}{5}$ (LCD, 10)

Step 2 Rename each numerator and each denominator with the LCD.

Step 3 Add the numerators.

$\frac{7}{10} + \frac{6}{10} = \frac{(7+6)}{10} = \frac{13}{10}$

$\frac{13}{10}$ inches total precipitation, or $1\frac{3}{10}$ inches.

Practice Problem On an electric bill, about $\frac{1}{8}$ of the energy is from solar energy and about $\frac{1}{10}$ is from wind power. How much of the total bill is from solar energy and wind power combined?

Example 3

In your body, $\frac{7}{10}$ of your muscle contractions are involuntary (cardiac and smooth muscle tissue). Smooth muscle makes $\frac{3}{15}$ of your muscle contractions. How many of your muscle contractions are made by cardiac muscle?

Step 1 Find the LCD of the fractions.

$\frac{7}{10} - \frac{3}{15}$ (LCD, 30)

Step 2 Rename each numerator and each denominator with the LCD.

$\frac{7 \times 3}{10 \times 3} = \frac{21}{30}$

$\frac{3 \times 2}{15 \times 2} = \frac{6}{30}$

Step 3 Subtract the numerators.

$\frac{21}{30} - \frac{6}{30} = \frac{(21-6)}{30} = \frac{15}{30}$

Step 4 Find the GCF.

$\frac{15}{30}$ (GCF, 15)

$\frac{1}{2}$

$\frac{1}{2}$ of all muscle contractions are cardiac muscle.

Example 4

Tony wants to make cookies that call for $\frac{3}{4}$ of a cup of flour, but he only has $\frac{1}{3}$ of a cup. How much more flour does he need?

Step 1 Find the LCD of the fractions.

$\frac{3}{4} - \frac{1}{3}$ (LCD, 12)

Step 2 Rename each numerator and each denominator with the LCD.

$\frac{3 \times 3}{4 \times 3} = \frac{9}{12}$

$\frac{1 \times 4}{3 \times 4} = \frac{4}{12}$

Step 3 Subtract the numerators.

$\frac{9}{12} - \frac{4}{12} = \frac{(9-4)}{12} = \frac{5}{12}$

$\frac{5}{12}$ of a cup of flour

Practice Problem Using the information provided to you in Example 3 above, determine how many muscle contractions are voluntary (skeletal muscle).

Math Skill Handbook • **SR-15**

Multiply Fractions To multiply with fractions, multiply the numerators and multiply the denominators. Find the simplest form if necessary.

Example

Multiply $\frac{3}{5}$ by $\frac{1}{3}$.

Step 1 Multiply the numerators and denominators.
$$\frac{3}{5} \times \frac{1}{3} = \frac{(3 \times 1)}{(5 \times 3)} = \frac{3}{15}$$

Step 2 Find the GCF.
$$\frac{3}{15} \text{ (GCF, 3)}$$

Step 3 Divide the numerator and denominator by the GCF.
$$\frac{3 \div 3}{15 \div 3} = \frac{1}{5}$$

$\frac{3}{5}$ multiplied by $\frac{1}{3}$ is $\frac{1}{5}$.

Practice Problem Multiply $\frac{3}{14}$ by $\frac{5}{16}$.

Find a Reciprocal Two numbers whose product is 1 are called multiplicative inverses, or reciprocals.

Example

Find the reciprocal of $\frac{3}{8}$.

Step 1 Inverse the fraction by putting the denominator on top and the numerator on the bottom.
$$\frac{8}{3}$$

The reciprocal of $\frac{3}{8}$ is $\frac{8}{3}$.

Practice Problem Find the reciprocal of $\frac{4}{9}$.

Divide Fractions To divide one fraction by another fraction, multiply the dividend by the reciprocal of the divisor. Find the simplest form if necessary.

Example 1

Divide $\frac{1}{9}$ by $\frac{1}{3}$.

Step 1 Find the reciprocal of the divisor. The reciprocal of $\frac{1}{3}$ is $\frac{3}{1}$.

Step 2 Multiply the dividend by the reciprocal of the divisor.
$$\frac{\frac{1}{9}}{\frac{1}{3}} = \frac{1}{9} \times \frac{3}{1} = \frac{(1 \times 3)}{(9 \times 1)} = \frac{3}{9}$$

Step 3 Find the GCF.
$$\frac{3}{9} \text{ (GCF, 3)}$$

Step 4 Divide the numerator and denominator by the GCF.
$$\frac{3 \div 3}{9 \div 3} = \frac{1}{3}$$

$\frac{1}{9}$ divided by $\frac{1}{3}$ is $\frac{1}{3}$.

Example 2

Divide $\frac{3}{5}$ by $\frac{1}{4}$.

Step 1 Find the reciprocal of the divisor. The reciprocal of $\frac{1}{4}$ is $\frac{4}{1}$.

Step 2 Multiply the dividend by the reciprocal of the divisor.
$$\frac{\frac{3}{5}}{\frac{1}{4}} = \frac{3}{5} \times \frac{4}{1} = \frac{(3 \times 4)}{(5 \times 1)} = \frac{12}{5}$$

$\frac{3}{5}$ divided by $\frac{1}{4}$ is $\frac{12}{5}$ or $2\frac{2}{5}$.

Practice Problem Divide $\frac{3}{11}$ by $\frac{7}{10}$.

Use Ratios

When you compare two numbers by division, you are using a ratio. Ratios can be written 3 to 5, 3:5, or $\frac{3}{5}$. Ratios, like fractions, also can be written in simplest form.

Ratios can represent one type of probability, called odds. This is a ratio that compares the number of ways a certain outcome occurs to the number of possible outcomes. For example, if you flip a coin 100 times, what are the odds that it will come up heads? There are two possible outcomes, heads or tails, so the odds of coming up heads are 50:100. Another way to say this is that 50 out of 100 times the coin will come up heads. In its simplest form, the ratio is 1:2.

Example 1

A chemical solution contains 40 g of salt and 64 g of baking soda. What is the ratio of salt to baking soda as a fraction in simplest form?

Step 1 Write the ratio as a fraction.

$$\frac{\text{salt}}{\text{baking soda}} = \frac{40}{64}$$

Step 2 Express the fraction in simplest form. The GCF of 40 and 64 is 8.

$$\frac{40}{64} = \frac{40 \div 8}{64 \div 8} = \frac{5}{8}$$

The ratio of salt to baking soda in the sample is 5:8.

Example 2

Sean rolls a 6-sided die 6 times. What are the odds that the side with a 3 will show?

Step 1 Write the ratio as a fraction.

$$\frac{\text{number of sides with a 3}}{\text{number of possible sides}} = \frac{1}{6}$$

Step 2 Multiply by the number of attempts.

$\frac{1}{6} \times 6$ attempts $= \frac{6}{6}$ attempts $= 1$ attempt

1 attempt out of 6 will show a 3.

Practice Problem Two metal rods measure 100 cm and 144 cm in length. What is the ratio of their lengths in simplest form?

Use Decimals

A fraction with a denominator that is a power of ten can be written as a decimal. For example, 0.27 means $\frac{27}{100}$. The decimal point separates the ones place from the tenths place.

Any fraction can be written as a decimal using division. For example, the fraction $\frac{5}{8}$ can be written as a decimal by dividing 5 by 8. Written as a decimal, it is 0.625.

Add or Subtract Decimals When adding and subtracting decimals, line up the decimal points before carrying out the operation.

Example 1

Find the sum of 47.68 and 7.80.

Step 1 Line up the decimal places when you write the numbers.

```
  47.68
+  7.80
```

Step 2 Add the decimals.

```
  ¹¹
  47.68
+  7.80
  55.48
```

The sum of 47.68 and 7.80 is 55.48.

Example 2

Find the difference of 42.17 and 15.85.

Step 1 Line up the decimal places when you write the number.

```
  42.17
- 15.85
```

Step 2 Subtract the decimals.

```
  ³¹¹
  4̷2̷.17
- 15.85
  26.32
```

The difference of 42.17 and 15.85 is 26.32.

Practice Problem Find the sum of 1.245 and 3.842.

Math Skill Handbook • **SR-17**

Multiply Decimals To multiply decimals, multiply the numbers like numbers without decimal points. Count the decimal places in each factor. The product will have the same number of decimal places as the sum of the decimal places in the factors.

> **Example**
>
> Multiply 2.4 by 5.9.
>
> **Step 1** Multiply the factors like two whole numbers.
>
> $24 \times 59 = 1416$
>
> **Step 2** Find the sum of the number of decimal places in the factors. Each factor has one decimal place, for a sum of two decimal places.
>
> **Step 3** The product will have two decimal places.
>
> 14.16
>
> The product of 2.4 and 5.9 is 14.16.

Practice Problem Multiply 4.6 by 2.2.

Divide Decimals When dividing decimals, change the divisor to a whole number. To do this, multiply both the divisor and the dividend by the same power of ten. Then place the decimal point in the quotient directly above the decimal point in the dividend. Then divide as you do with whole numbers.

> **Example**
>
> Divide 8.84 by 3.4.
>
> **Step 1** Multiply both factors by 10.
>
> $3.4 \times 10 = 34, 8.84 \times 10 = 88.4$
>
> **Step 2** Divide 88.4 by 34.
>
> $$\begin{array}{r} 2.6 \\ 34\overline{)88.4} \\ -68 \\ \hline 204 \\ -204 \\ \hline 0 \end{array}$$
>
> 8.84 divided by 3.4 is 2.6.

Practice Problem Divide 75.6 by 3.6.

Use Proportions

An equation that shows that two ratios are equivalent is a proportion. The ratios $\frac{2}{4}$ and $\frac{5}{10}$ are equivalent, so they can be written as $\frac{2}{4} = \frac{5}{10}$. This equation is a proportion.

When two ratios form a proportion, the cross products are equal. To find the cross products in the proportion $\frac{2}{4} = \frac{5}{10}$, multiply the 2 and the 10, and the 4 and the 5. Therefore $2 \times 10 = 4 \times 5$, or $20 = 20$.

Because you know that both ratios are equal, you can use cross products to find a missing term in a proportion. This is known as solving the proportion.

> **Example**
>
> The heights of a tree and a pole are proportional to the lengths of their shadows. The tree casts a shadow of 24 m when a 6-m pole casts a shadow of 4 m. What is the height of the tree?
>
> **Step 1** Write a proportion.
>
> $$\frac{\text{height of tree}}{\text{height of pole}} = \frac{\text{length of tree's shadow}}{\text{length of pole's shadow}}$$
>
> **Step 2** Substitute the known values into the proportion. Let h represent the unknown value, the height of the tree.
>
> $\frac{h}{6} \times \frac{24}{4}$
>
> **Step 3** Find the cross products.
>
> $h \times 4 = 6 \times 24$
>
> **Step 4** Simplify the equation.
>
> $4h \times 144$
>
> **Step 5** Divide each side by 4.
>
> $\frac{4h}{4} \times \frac{144}{4}$
>
> $h = 36$
>
> The height of the tree is 36 m.

Practice Problem The ratios of the weights of two objects on the Moon and on Earth are in proportion. A rock weighing 3 N on the Moon weighs 18 N on Earth. How much would a rock that weighs 5 N on the Moon weigh on Earth?

Use Percentages

The word *percent* means "out of one hundred." It is a ratio that compares a number to 100. Suppose you read that 77 percent of Earth's surface is covered by water. That is the same as reading that the fraction of Earth's surface covered by water is $\frac{77}{100}$. To express a fraction as a percent, first find the equivalent decimal for the fraction. Then, multiply the decimal by 100 and add the percent symbol.

Example 1

Express $\frac{13}{20}$ as a percent.

Step 1 Find the equivalent decimal for the fraction.

```
      0.65
20)13.00
    12 0
     1 00
     1 00
        0
```

Step 2 Rewrite the fraction $\frac{13}{20}$ as 0.65.

Step 3 Multiply 0.65 by 100 and add the % symbol.

$0.65 \times 100 = 65 = 65\%$

So, $\frac{13}{20} = 65\%$.

This also can be solved as a proportion.

Example 2

Express $\frac{13}{20}$ as a percent.

Step 1 Write a proportion.

$\frac{13}{20} = \frac{x}{100}$

Step 2 Find the cross products.

$1300 = 20x$

Step 3 Divide each side by 20.

$\frac{1300}{20} = \frac{20x}{20}$

$65\% = x$

Practice Problem In one year, 73 of 365 days were rainy in one city. What percent of the days in that city were rainy?

Solve One-Step Equations

A statement that two expressions are equal is an equation. For example, $A = B$ is an equation that states that A is equal to B.

An equation is solved when a variable is replaced with a value that makes both sides of the equation equal. To make both sides equal the inverse operation is used. Addition and subtraction are inverses, and multiplication and division are inverses.

Example 1

Solve the equation $x - 10 = 35$.

Step 1 Find the solution by adding 10 to each side of the equation.

$x - 10 = 35$
$x - 10 + 10 = 35 - 10$
$x = 45$

Step 2 Check the solution.

$x - 10 = 35$
$45 - 10 = 35$
$35 = 35$

Both sides of the equation are equal, so $x = 45$.

Example 2

In the formula $a = bc$, find the value of c if $a = 20$ and $b = 2$.

Step 1 Rearrange the formula so the unknown value is by itself on one side of the equation by dividing both sides by b.

$a = bc$
$\frac{a}{b} = \frac{bc}{b}$
$\frac{a}{b} = c$

Step 2 Replace the variables a and b with the values that are given.

$\frac{a}{b} = c$
$\frac{20}{2} = c$
$10 = c$

Step 3 Check the solution.

$a = bc$
$20 = 2 \times 10$
$20 = 20$

Both sides of the equation are equal, so $c = 10$ is the solution when $a = 20$ and $b = 2$.

Practice Problem In the formula $h = gd$, find the value of d if $g = 12.3$ and $h = 17.4$.

Use Statistics

The branch of mathematics that deals with collecting, analyzing, and presenting data is statistics. In statistics, there are three common ways to summarize data with a single number—the mean, the median, and the mode.

The **mean** of a set of data is the arithmetic average. It is found by adding the numbers in the data set and dividing by the number of items in the set.

The **median** is the middle number in a set of data when the data are arranged in numerical order. If there were an even number of data points, the median would be the mean of the two middle numbers.

The **mode** of a set of data is the number or item that appears most often.

Another number that often is used to describe a set of data is the range. The **range** is the difference between the largest number and the smallest number in a set of data.

Example

The speeds (in m/s) for a race car during five different time trials are 39, 37, 44, 36, and 44.

To find the mean:

Step 1 Find the sum of the numbers.

$39 + 37 + 44 + 36 + 44 = 200$

Step 2 Divide the sum by the number of items, which is 5.

$200 \div 5 = 40$

The mean is 40 m/s.

To find the median:

Step 1 Arrange the measures from least to greatest.

36, 37, 39, 44, 44

Step 2 Determine the middle measure.

36, 37, <u>39</u>, 44, 44

The median is 39 m/s.

To find the mode:

Step 1 Group the numbers that are the same together.

44, 44, 36, 37, 39

Step 2 Determine the number that occurs most in the set.

<u>44, 44</u>, 36, 37, 39

The mode is 44 m/s.

To find the range:

Step 1 Arrange the measures from greatest to least.

44, 44, 39, 37, 36

Step 2 Determine the greatest and least measures in the set.

<u>44</u>, 44, 39, 37, <u>36</u>

Step 3 Find the difference between the greatest and least measures.

$44 - 36 = 8$

The range is 8 m/s.

Practice Problem Find the mean, median, mode, and range for the data set 8, 4, 12, 8, 11, 14, 16.

A **frequency table** shows how many times each piece of data occurs, usually in a survey. **Table 1** below shows the results of a student survey on favorite color.

Table 1 Student Color Choice		
Color	Tally	Frequency
red	IIII	4
blue	TH̶L	5
black	II	2
green	III	3
purple	TH̶L II	7
yellow	TH̶L I	6

Based on the frequency table data, which color is the favorite?

SR-20 • Math Skill Handbook

Use Geometry

The branch of mathematics that deals with the measurement, properties, and relationships of points, lines, angles, surfaces, and solids is called geometry.

Perimeter The **perimeter** (P) is the distance around a geometric figure. To find the perimeter of a rectangle, add the length and width and multiply that sum by two, or $2(l + w)$. To find perimeters of irregular figures, add the length of the sides.

Example 1

Find the perimeter of a rectangle that is 3 m long and 5 m wide.

Step 1 You know that the perimeter is 2 times the sum of the width and length.

$P = 2(3 \text{ m} + 5 \text{ m})$

Step 2 Find the sum of the width and length.

$P = 2(8 \text{ m})$

Step 3 Multiply by 2.

$P = 16 \text{ m}$

The perimeter is 16 m.

Example 2

Find the perimeter of a shape with sides measuring 2 cm, 5 cm, 6 cm, 3 cm.

Step 1 You know that the perimeter is the sum of all the sides.

$P = 2 + 5 + 6 + 3$

Step 2 Find the sum of the sides.

$P = 2 + 5 + 6 + 3$

$P = 16$

The perimeter is 16 cm.

Practice Problem Find the perimeter of a rectangle with a length of 18 m and a width of 7 m.

Practice Problem Find the perimeter of a triangle measuring 1.6 cm by 2.4 cm by 2.4 cm.

Area of a Rectangle The **area** (A) is the number of square units needed to cover a surface. To find the area of a rectangle, multiply the length times the width, or $l \times w$. When finding area, the units also are multiplied. Area is given in square units.

Example

Find the area of a rectangle with a length of 1 cm and a width of 10 cm.

Step 1 You know that the area is the length multiplied by the width.

$A = (1 \text{ cm} \times 10 \text{ cm})$

Step 2 Multiply the length by the width. Also multiply the units.

$A = 10 \text{ cm}^2$

The area is 10 cm².

Practice Problem Find the area of a square whose sides measure 4 m.

Area of a Triangle To find the area of a triangle, use the formula:

$A = \frac{1}{2}(\text{base} \times \text{height})$

The base of a triangle can be any of its sides. The height is the perpendicular distance from a base to the opposite endpoint, or vertex.

Example

Find the area of a triangle with a base of 18 m and a height of 7 m.

Step 1 You know that the area is $\frac{1}{2}$ the base times the height.

$A = \frac{1}{2}(18 \text{ m} \times 7 \text{ m})$

Step 2 Multiply $\frac{1}{2}$ by the product of 18×7. Multiply the units.

$A = \frac{1}{2}(126 \text{ m}^2)$

$A = 63 \text{ m}^2$

The area is 63 m².

Practice Problem Find the area of a triangle with a base of 27 cm and a height of 17 cm.

Math Skill Handbook • **SR-21**

Circumference of a Circle The **diameter** (*d*) of a circle is the distance across the circle through its center, and the **radius** (r) is the distance from the center to any point on the circle. The radius is half of the diameter. The distance around the circle is called the **circumference** (C). The formula for finding the circumference is:

$C = 2\pi r$ or $C = \pi d$

The circumference divided by the diameter is always equal to 3.1415926… This nonterminating and nonrepeating number is represented by the Greek letter π (pi). An approximation often used for π is 3.14.

Example 1

Find the circumference of a circle with a radius of 3 m.

Step 1 You know the formula for the circumference is 2 times the radius times π.

$C = 2\pi(3)$

Step 2 Multiply 2 times the radius.

$C = 6\pi$

Step 3 Multiply by π.

$C \approx 19$ m

The circumference is about 19 m.

Example 2

Find the circumference of a circle with a diameter of 24.0 cm.

Step 1 You know the formula for the circumference is the diameter times π.

$C = \pi(24.0)$

Step 2 Multiply the diameter by π.

$C \approx 75.4$ cm

The circumference is about 75.4 cm.

Practice Problem Find the circumference of a circle with a radius of 19 cm.

Area of a Circle The formula for the area of a circle is: $A = \pi r^2$

Example 1

Find the area of a circle with a radius of 4.0 cm.

Step 1 $A = \pi(4.0)^2$

Step 2 Find the square of the radius.

$A = 16\pi$

Step 3 Multiply the square of the radius by π.

$A \approx 50$ cm^2

The area of the circle is about 50 cm^2.

Example 2

Find the area of a circle with a radius of 225 m.

Step 1 $A = \pi(225)^2$

Step 2 Find the square of the radius.

$A = 50625\pi$

Step 3 Multiply the square of the radius by π.

$A \approx 159043.1$

The area of the circle is about 159043.1 m^2.

Example 3

Find the area of a circle whose diameter is 20.0 mm.

Step 1 Remember that the radius is half of the diameter.

$A = \pi\left(\dfrac{20.0}{2}\right)^2$

Step 2 Find the radius.

$A = \pi(10.0)^2$

Step 3 Find the square of the radius.

$A = 100\pi$

Step 4 Multiply the square of the radius by π.

$A \approx 314$ mm^2

The area of the circle is about 314 mm^2.

Practice Problem Find the area of a circle with a radius of 16 m.

Volume The measure of space occupied by a solid is the **volume** (V). To find the volume of a rectangular solid multiply the length times width times height, or $V = l \times w \times h$. It is measured in cubic units, such as cubic centimeters (cm^3).

Example

Find the volume of a rectangular solid with a length of 2.0 m, a width of 4.0 m, and a height of 3.0 m.

Step 1 You know the formula for volume is the length times the width times the height.

$V = 2.0 \text{ m} \times 4.0 \text{ m} \times 3.0 \text{ m}$

Step 2 Multiply the length times the width times the height.

$V = 24 \text{ m}^3$

The volume is 24 m^3.

Practice Problem Find the volume of a rectangular solid that is 8 m long, 4 m wide, and 4 m high.

To find the volume of other solids, multiply the area of the base times the height.

Example 1

Find the volume of a solid that has a triangular base with a length of 8.0 m and a height of 7.0 m. The height of the entire solid is 15.0 m.

Step 1 You know that the base is a triangle, and the area of a triangle is $\frac{1}{2}$ the base times the height, and the volume is the area of the base times the height.

$V = \left[\frac{1}{2}(b \times h)\right] \times 15$

Step 2 Find the area of the base.

$V = \left[\frac{1}{2}(8 \times 7)\right] \times 15$

$V = \left(\frac{1}{2} \times 56\right) \times 15$

Step 3 Multiply the area of the base by the height of the solid.

$V = 28 \times 15$

$V = 420 \text{ m}^3$

The volume is 420 m^3.

Example 2

Find the volume of a cylinder that has a base with a radius of 12.0 cm, and a height of 21.0 cm.

Step 1 You know that the base is a circle, and the area of a circle is the square of the radius times π, and the volume is the area of the base times the height.

$V = (\pi r^2) \times 21$

$V = (\pi 12^2) \times 21$

Step 2 Find the area of the base.

$V = 144\pi \times 21$

$V = 452 \times 21$

Step 3 Multiply the area of the base by the height of the solid.

$V \approx 9{,}500 \text{ cm}^3$

The volume is about 9,500 cm^3.

Example 3

Find the volume of a cylinder that has a diameter of 15 mm and a height of 4.8 mm.

Step 1 You know that the base is a circle with an area equal to the square of the radius times π. The radius is one-half the diameter. The volume is the area of the base times the height.

$V = (\pi r^2) \times 4.8$

$V = \left[\pi \left(\frac{1}{2} \times 15\right)^2\right] \times 4.8$

$V = (\pi 7.5^2) \times 4.8$

Step 2 Find the area of the base.

$V = 56.25\pi \times 4.8$

$V \approx 176.71 \times 4.8$

Step 3 Multiply the area of the base by the height of the solid.

$V \approx 848.2$

The volume is about 848.2 mm^3.

Practice Problem Find the volume of a cylinder with a diameter of 7 cm in the base and a height of 16 cm.

Science Applications

Measure in SI

The metric system of measurement was developed in 1795. A modern form of the metric system, called the International System (SI), was adopted in 1960 and provides the standard measurements that all scientists around the world can understand.

The SI system is convenient because unit sizes vary by powers of 10. Prefixes are used to name units. Look at **Table 2** for some common SI prefixes and their meanings.

Table 2 Common SI Prefixes

Prefix	Symbol	Meaning	
kilo–	k	1,000	thousandth
hecto–	h	100	hundred
deka–	da	10	ten
deci–	d	0.1	tenth
centi–	c	0.01	hundreth
milli–	m	0.001	thousandth

Example

How many grams equal one kilogram?

Step 1 Find the prefix *kilo–* in **Table 2**.

Step 2 Using **Table 2**, determine the meaning of *kilo–*. According to the table, it means 1,000. When the prefix *kilo–* is added to a unit, it means that there are 1,000 of the units in a "kilounit."

Step 3 Apply the prefix to the units in the question. The units in the question are grams. There are 1,000 grams in a kilogram.

Practice Problem Is a milligram larger or smaller than a gram? How many of the smaller units equal one larger unit? What fraction of the larger unit does one smaller unit represent?

Dimensional Analysis

Convert SI Units In science, quantities such as length, mass, and time sometimes are measured using different units. A process called dimensional analysis can be used to change one unit of measure to another. This process involves multiplying your starting quantity and units by one or more conversion factors. A conversion factor is a ratio equal to one and can be made from any two equal quantities with different units. If 1,000 mL equal 1 L then two ratios can be made.

$$\frac{1,000 \text{ mL}}{1 \text{ L}} = \frac{1 \text{ L}}{1,000 \text{ mL}} = 1$$

One can convert between units in the SI system by using the equivalents in **Table 2** to make conversion factors.

Example

How many cm are in 4 m?

Step 1 Write conversion factors for the units given. From **Table 2**, you know that 100 cm = 1 m. The conversion factors are

$$\frac{100 \text{ cm}}{1 \text{ m}} \text{ and } \frac{1 \text{ m}}{100 \text{ cm}}$$

Step 2 Decide which conversion factor to use. Select the factor that has the units you are converting from (m) in the denominator and the units you are converting to (cm) in the numerator.

$$\frac{100 \text{ cm}}{1 \text{ m}}$$

Step 3 Multiply the starting quantity and units by the conversion factor. Cancel the starting units with the units in the denominator. There are 400 cm in 4 m.

$$4 \text{ m} = \frac{100 \text{ cm}}{1 \text{ m}} = 400 \text{ cm}$$

Practice Problem How many milligrams are in one kilogram? (Hint: You will need to use two conversion factors from **Table 2**.)

Table 3 Unit System Equivalents

Type of Measurement	Equivalent
Length	1 in = 2.54 cm 1 yd = 0.91 m 1 mi = 1.61 km
Mass and weight*	1 oz = 28.35 g 1 lb = 0.45 kg 1 ton (short) = 0.91 tonnes (metric tons) 1 lb = 4.45 N
Volume	1 in^3 = 16.39 cm^3 1 qt = 0.95 L 1 gal = 3.78 L
Area	1 in^2 = 6.45 cm^2 1 yd^2 = 0.83 m^2 1 mi^2 = 2.59 km^2 1 acre = 0.40 hectares
Temperature	$°C = \frac{(°F - 32)}{1.8}$ $K = °C + 273$

*Weight is measured in standard Earth gravity.

Convert Between Unit Systems Table 3 gives a list of equivalents that can be used to convert between English and SI units.

Example

If a meterstick has a length of 100 cm, how long is the meterstick in inches?

Step 1 Write the conversion factors for the units given. From **Table 3,** 1 in = 2.54 cm.

$$\frac{1 \text{ in}}{2.54 \text{ cm}} \text{ and } \frac{2.54 \text{ cm}}{1 \text{ in}}$$

Step 2 Determine which conversion factor to use. You are converting from cm to in. Use the conversion factor with cm on the bottom.

$$\frac{1 \text{ in}}{2.54 \text{ cm}}$$

Step 3 Multiply the starting quantity and units by the conversion factor. Cancel the starting units with the units in the denominator. Round your answer to the nearest tenth.

$$100 \text{ cm} \times \frac{1 \text{ in}}{2.54 \text{ cm}} = 39.37 \text{ in}$$

The meterstick is about 39.4 in long.

Practice Problem 1 A book has a mass of 5 lb. What is the mass of the book in kg?

Practice Problem 2 Use the equivalent for in and cm (1 in = 2.54 cm) to show how 1 in^3 ≈ 16.39 cm^3.

Precision and Significant Digits

When you make a measurement, the value you record depends on the precision of the measuring instrument. This precision is represented by the number of significant digits recorded in the measurement. When counting the number of significant digits, all digits are counted except zeros at the end of a number with no decimal point such as 2,050, and zeros at the beginning of a decimal such as 0.03020. When adding or subtracting numbers with different precision, round the answer to the smallest number of decimal places of any number in the sum or difference. When multiplying or dividing, the answer is rounded to the smallest number of significant digits of any number being multiplied or divided.

Example

The lengths 5.28 and 5.2 are measured in meters. Find the sum of these lengths and record your answer using the correct number of significant digits.

Step 1 Find the sum.

    5.28 m   2 digits after the decimal
  + 5.2 m   1 digit after the decimal
   10.48 m

Step 2 Round to one digit after the decimal because the least number of digits after the decimal of the numbers being added is 1.

The sum is 10.5 m.

Practice Problem 1 How many significant digits are in the measurement 7,071,301 m? How many significant digits are in the measurement 0.003010 g?

Practice Problem 2 Multiply 5.28 and 5.2 using the rule for multiplying and dividing. Record the answer using the correct number of significant digits.

Scientific Notation

Many times numbers used in science are very small or very large. Because these numbers are difficult to work with scientists use scientific notation. To write numbers in scientific notation, move the decimal point until only one non-zero digit remains on the left. Then count the number of places you moved the decimal point and use that number as a power of ten. For example, the average distance from the Sun to Mars is 227,800,000,000 m. In scientific notation, this distance is 2.278×10^{11} m. Because you moved the decimal point to the left, the number is a positive power of ten.

The mass of an electron is about 0.000 000 000 000 000 000 000 000 000 000 911 kg. Expressed in scientific notation, this mass is 9.11×10^{-31} kg. Because the decimal point was moved to the right, the number is a negative power of ten.

Example

Earth is 149,600,000 km from the Sun. Express this in scientific notation.

Step 1 Move the decimal point until one non-zero digit remains on the left.

1.496 000 00

Step 2 Count the number of decimal places you have moved. In this case, eight.

Step 2 Show that number as a power of ten, 10^8.

Earth is 1.496×10^8 km from the Sun.

Practice Problem 1 How many significant digits are in 149,600,000 km? How many significant digits are in 1.496×10^8 km?

Practice Problem 2 Parts used in a high performance car must be measured to 7×10^{-6} m. Express this number as a decimal.

Practice Problem 3 A CD is spinning at 539 revolutions per minute. Express this number in scientific notation.

Make and Use Graphs

Data in tables can be displayed in a graph—a visual representation of data. Common graph types include line graphs, bar graphs, and circle graphs.

Line Graph A line graph shows a relationship between two variables that change continuously. The independent variable is changed and is plotted on the x-axis. The dependent variable is observed, and is plotted on the y-axis.

Example

Draw a line graph of the data below from a cyclist in a long-distance race.

Table 4 Bicycle Race Data

Time (h)	Distance (km)
0	0
1	8
2	16
3	24
4	32
5	40

Step 1 Determine the x-axis and y-axis variables. Time varies independently of distance and is plotted on the x-axis. Distance is dependent on time and is plotted on the y-axis.

Step 2 Determine the scale of each axis. The x-axis data ranges from 0 to 5. The y-axis data ranges from 0 to 50.

Step 3 Using graph paper, draw and label the axes. Include units in the labels.

Step 4 Draw a point at the intersection of the time value on the x-axis and corresponding distance value on the y-axis. Connect the points and label the graph with a title, as shown in **Figure 8**.

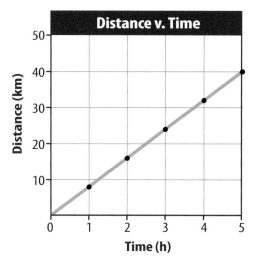

Figure 8 This line graph shows the relationship between distance and time during a bicycle ride.

Practice Problem A puppy's shoulder height is measured during the first year of her life. The following measurements were collected: (3 mo, 52 cm), (6 mo, 72 cm), (9 mo, 83 cm), (12 mo, 86 cm). Graph this data.

Find a Slope The slope of a straight line is the ratio of the vertical change, rise, to the horizontal change, run.

$$\text{Slope} = \frac{\text{vertical change (rise)}}{\text{horizontal change (run)}} = \frac{\text{change in } y}{\text{change in } x}$$

Example

Find the slope of the graph in **Figure 8**.

Step 1 You know that the slope is the change in y divided by the change in x.

$$\text{Slope} = \frac{\text{change in } y}{\text{change in } x}$$

Step 2 Determine the data points you will be using. For a straight line, choose the two sets of points that are the farthest apart.

$$\text{Slope} = \frac{(40 - 0) \text{ km}}{(5 - 0) \text{ h}}$$

Step 3 Find the change in y and x.

$$\text{Slope} = \frac{40 \text{ km}}{5 \text{ h}}$$

Step 4 Divide the change in y by the change in x.

$$\text{Slope} = \frac{8 \text{ km}}{\text{h}}$$

The slope of the graph is 8 km/h.

Bar Graph To compare data that does not change continuously you might choose a bar graph. A bar graph uses bars to show the relationships between variables. The *x*-axis variable is divided into parts. The parts can be numbers such as years, or a category such as a type of animal. The *y*-axis is a number and increases continuously along the axis.

Example

A recycling center collects 4.0 kg of aluminum on Monday, 1.0 kg on Wednesday, and 2.0 kg on Friday. Create a bar graph of this data.

Step 1 Select the *x*-axis and *y*-axis variables. The measured numbers (the masses of aluminum) should be placed on the *y*-axis. The variable divided into parts (collection days) is placed on the *x*-axis.

Step 2 Create a graph grid like you would for a line graph. Include labels and units.

Step 3 For each measured number, draw a vertical bar above the *x*-axis value up to the *y*-axis value. For the first data point, draw a vertical bar above Monday up to 4.0 kg.

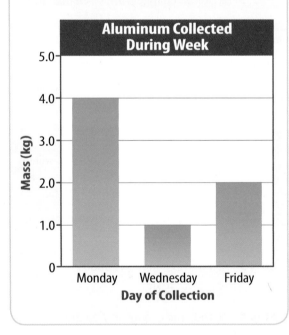

Practice Problem Draw a bar graph of the gases in air: 78% nitrogen, 21% oxygen, 1% other gases.

Circle Graph To display data as parts of a whole, you might use a circle graph. A circle graph is a circle divided into sections that represent the relative size of each piece of data. The entire circle represents 100%, half represents 50%, and so on.

Example

Air is made up of 78% nitrogen, 21% oxygen, and 1% other gases. Display the composition of air in a circle graph.

Step 1 Multiply each percent by 360° and divide by 100 to find the angle of each section in the circle.

$$78\% \times \frac{360°}{100} = 280.8°$$

$$21\% \times \frac{360°}{100} = 75.6°$$

$$1\% \times \frac{360°}{100} = 3.6°$$

Step 2 Use a compass to draw a circle and to mark the center of the circle. Draw a straight line from the center to the edge of the circle.

Step 3 Use a protractor and the angles you calculated to divide the circle into parts. Place the center of the protractor over the center of the circle and line the base of the protractor over the straight line.

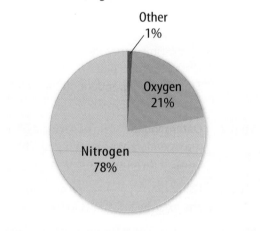

Practice Problem Draw a circle graph to represent the amount of aluminum collected during the week shown in the bar graph to the left.

Student Study Guides & Instructions
By Dinah Zike

1. You will find suggestions for Study Guides, also known as Foldables or books, in each chapter lesson and as a final project. Look at the end of the chapter to determine the project format and glue the Foldables in place as you progress through the chapter lessons.

2. Creating the Foldables or books is simple and easy to do by using copy paper, art paper, and internet printouts. Photocopies of maps, diagrams, or your own illustrations may also be used for some of the Foldables. Notebook paper is the most common source of material for study guides and 83% of all Foldables are created from it. When folded to make books, notebook paper Foldables easily fit into 11" × 17" or 12" × 18" chapter projects with space left over. Foldables made using photocopy paper are slightly larger and they fit into Projects, but snugly. Use the least amount of glue, tape, and staples needed to assemble the Foldables.

3. Seven of the Foldables can be made using either small or large paper. When 11" × 17" or 12" × 18" paper is used, these become projects for housing smaller Foldables. Project format boxes are located within the instructions to remind you of this option.

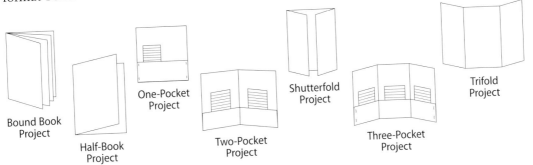

4. Use one-gallon self-locking plastic bags to store your projects. Place strips of two-inch clear tape along the left, long side of the bag and punch holes through the taped edge. Cut the bottom corners off the bag so it will not hold air. Store this Project Portfolio inside a three-hole binder. To store a large collection of project bags, use a giant laundry-soap box. Holes can be punched in some of the Foldable Projects so they can be stored in a three-hole binder without using a plastic bag. Punch holes in the pocket books before gluing or stapling the pocket.

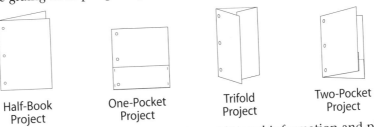

5. Maximize the use of the projects by collecting additional information and placing it on the back of the project and other unused spaces of the large Foldables.

Half-Book Foldable® By Dinah Zike

Step 1 Fold a sheet of notebook or copy paper in half.

Label the exterior tab and use the inside space to write information.

PROJECT FORMAT
Use 11" × 17" or 12" × 18" paper on the horizontal axis to make a large project book.

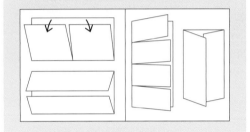

Variations
Paper can be folded horizontally, like a *hamburger* or vertically, like a *hot dog*.

A

B

C Half-books can be folded so that one side is ½ inch longer than the other side. A title or question can be written on the extended tab.

- -

Worksheet Foldable or Folded Book® By Dinah Zike

Step 1 Make a half-book (see above) using work sheets, internet print-outs, diagrams, or maps.

Step 2 Fold it in half again.

Variations

A This folded sheet as a small book with two pages can be used for comparing and contrasting, cause and effect, or other skills.

B When the sheet of paper is open, the four sections can be used separately or used collectively to show sequences or steps.

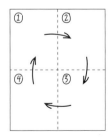

SR-30 • Foldables Handbook

Two-Tab and Concept-Map Foldable® By Dinah Zike

Step 1 Fold a sheet of notebook or copy paper in half vertically or horizontally.

Step 2 Fold it in half again, as shown.

Step 3 Unfold once and cut along the fold line or valley of the top flap to make two flaps.

Variations

A Concept maps can be made by leaving a ½ inch tab at the top when folding the paper in half. Use arrows and labels to relate topics to the primary concept.

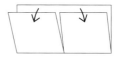

B Use two sheets of paper to make multiple page tab books. Glue or staple books together at the top fold.

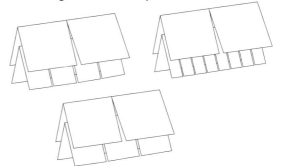

Three-Quarter Foldable® By Dinah Zike

Step 1 Make a two-tab book (see above) and cut the left tab off at the top of the fold line.

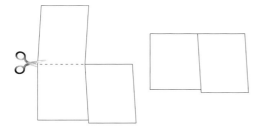

Variations

A Use this book to draw a diagram or a map on the exposed left tab. Write questions about the illustration on the top right tab and provide complete answers on the space under the tab.

B Compose a self-test using multiple choice answers for your questions. Include the correct answer with three wrong responses. The correct answers can be written on the back of the book or upside down on the bottom of the inside page.

Three-Tab Foldable® By Dinah Zike

Step 1 Fold a sheet of paper in half horizontally.

Step 2 Fold into thirds.

Step 3 Unfold and cut along the folds of the top flap to make three sections.

Variations

A Before cutting the three tabs draw a Venn diagram across the front of the book.

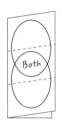

B Make a space to use for titles or concept maps by leaving a ½ inch tab at the top when folding the paper in half.

Four-Tab Foldable® By Dinah Zike

Step 1 Fold a sheet of paper in half horizontally.

Step 2 Fold in half and then fold each half as shown below.

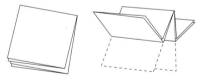

Step 3 Unfold and cut along the fold lines of the top flap to make four tabs.

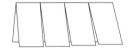

Variations

A Make a space to use for titles or concept maps by leaving a ½ inch tab at the top when folding the paper in half.

B Use the book on the vertical axis, with or without an extended tab.

Folding Fifths for a Foldable® By Dinah Zike

Step 1 Fold a sheet of paper in half horizontally.

Step 2 Fold again so one-third of the paper is exposed and two-thirds are covered.

Step 3 Fold the two-thirds section in half.

Step 4 Fold the one-third section, a single thickness, backward to make a fold line.

Variations

A Unfold and cut along the fold lines to make five tabs.

B Make a five-tab book with a ½ inch tab at the top (see two-tab instructions).

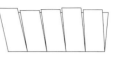

C Use 11" × 17" or 12" × 18" paper and fold into fifths for a five-column and/or row table or chart.

- -

Folded Table or Chart, and Trifold Foldable® By Dinah Zike

Step 1 Fold a sheet of paper in the required number of vertical columns for the table or chart.

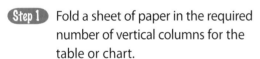

Step 2 Fold the horizontal rows needed for the table or chart.

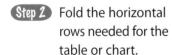

PROJECT FORMAT
Use 11" × 17" or 12" × 18" paper and fold it to make a large trifold project book or larger tables and charts.

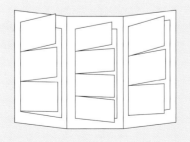

Variations

A Make a trifold by folding the paper into thirds vertically or horizontally.

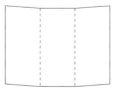

B Make a trifold book. Unfold it and draw a Venn diagram on the inside.

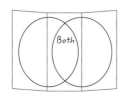

Foldables Handbook • **SR-33**

Two or Three-Pockets Foldable® By Dinah Zike

Step 1 Fold up the long side of a horizontal sheet of paper about 5 cm.

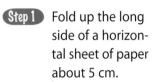

Step 2 Fold the paper in half.

Step 3 Open the paper and glue or staple the outer edges to make two compartments.

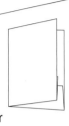

Variations

A Make a multi-page booklet by gluing several pocket books together.

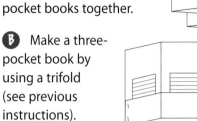

B Make a three-pocket book by using a trifold (see previous instructions).

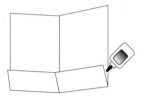

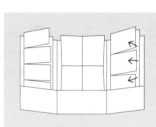

PROJECT FORMAT
Use 11" × 17" or 12" × 18" paper and fold it horizontally to make a large multi-pocket project.

Matchbook Foldable® By Dinah Zike

Step 1 Fold a sheet of paper almost in half and make the back edge about 1–2 cm longer than the front edge.

Step 2 Find the midpoint of the shorter flap.

Step 3 Open the paper and cut the short side along the midpoint making two tabs.

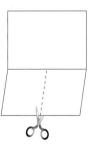

Step 4 Close the book and fold the tab over the short side.

Variations

A Make a single-tab matchbook by skipping Steps 2 and 3.

B Make two smaller matchbooks by cutting the single-tab matchbook in half.

Shutterfold Foldable® By Dinah Zike

Step 1 Begin as if you were folding a vertical sheet of paper in half, but instead of creasing the paper, pinch it to show the midpoint.

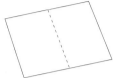

Step 2 Fold the top and bottom to the middle and crease the folds.

Variations

A Use the shutterfold on the horizontal axis.

B Create a center tab by leaving .5–2 cm between the flaps in Step 2.

PROJECT FORMAT
Use 11" × 17" or 12" × 18" paper and fold it to make a large shutterfold project.

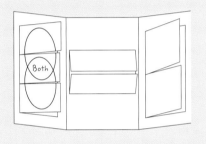

Four-Door Foldable® By Dinah Zike

Step 1 Make a shutterfold (see above).

Step 2 Fold the sheet of paper in half.

Step 3 Open the last fold and cut along the inside fold lines to make four tabs.

Variations

A Use the four-door book on the opposite axis.

B Create a center tab by leaving .5–2 cm between the flaps in Step 1.

Foldables Handbook • **SR-35**

Bound Book Foldable® By Dinah Zike

Step 1 Fold three sheets of paper in half. Place the papers in a stack, leaving about .5 cm between each top fold. Mark all three sheets about 3 cm from the outer edges.

Step 2 Using two of the sheets, cut from the outer edges to the marked spots on each side. On the other sheet, cut between the marked spots.

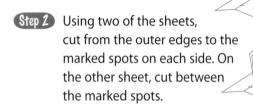

Step 3 Take the two sheets from Step 1 and slide them through the cut in the third sheet to make a 12-page book.

Step 4 Fold the bound pages in half to form a book.

Variation

A Use two sheets of paper to make an eight-page book, or increase the number of pages by using more than three sheets.

PROJECT FORMAT
Use two or more sheets of 11" × 17" or 12" × 18" paper and fold it to make a large bound book project.

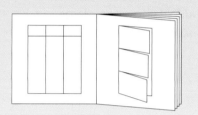

Accordian Foldable® By Dinah Zike

Step 1 Fold the selected paper in half vertically, like a *hamburger*.

Step 2 Cut each sheet of folded paper in half along the fold lines.

Step 3 Fold each half-sheet almost in half, leaving a 2 cm tab at the top.

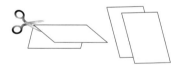

Step 4 Fold the top tab over the short side, then fold it in the opposite direction.

Variations

A Glue the straight edge of one paper inside the tab of another sheet. Leave a tab at the end of the book to add more pages.

B Tape the straight edge of one paper to the tab of another sheet, or just tape the straight edges of nonfolded paper end to end to make an accordian.

C Use whole sheets of paper to make a large accordian.

Layered Foldable® By Dinah Zike

Step 1 Stack two sheets of paper about 1–2 cm apart. Keep the right and left edges even.

Step 2 Fold up the bottom edges to form four tabs. Crease the fold to hold the tabs in place.

Step 3 Staple along the folded edge, or open and glue the papers together at the fold line.

Variations

A Rotate the book so the fold is at the top or to the side.

B Extend the book by using more than two sheets of paper.

Envelope Foldable® By Dinah Zike

Step 1 Fold a sheet of paper into a *taco*. Cut off the tab at the top.

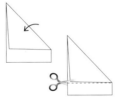

Step 2 Open the *taco* and fold it the opposite way making another *taco* and an X-fold pattern on the sheet of paper.

Step 3 Cut a map, illustration, or diagram to fit the inside of the envelope.

Step 4 Use the outside tabs for labels and inside tabs for writing information.

Variations

A Use 11″ × 17″ or 12″ × 18″ paper to make a large envelope.

B Cut off the points of the four tabs to make a window in the middle of the book.

Sentence Strip Foldable® By Dinah Zike

Step 1 Fold two sheets of paper in half vertically, like a *hamburger*.

Step 2 Unfold and cut along fold lines making four half sheets.

Step 3 Fold each half sheet in half horizontally, like a *hot dog*.

Step 4 Stack folded horizontal sheets evenly and staple together on the left side.

Step 5 Open the top flap of the first sentence strip and make a cut about 2 cm from the stapled edge to the fold line. This forms a flap that can be raisied and lowered. Repeat this step for each sentence strip.

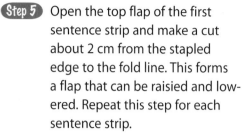

Variations

A Expand this book by using more than two sheets of paper.

B Use whole sheets of paper to make large books.

Pyramid Foldable® By Dinah Zike

Step 1 Fold a sheet of paper into a *taco*. Crease the fold line, but do not cut it off.

Step 2 Open the folded sheet and refold it like a *taco* in the opposite direction to create an X-fold pattern.

Step 3 Cut one fold line as shown, stopping at the center of the X-fold to make a flap.

Step 4 Outline the fold lines of the X-fold. Label the three front sections and use the inside spaces for notes. Use the tab for the title.

Step 5 Glue the tab into a project book or notebook. Use the space under the pyramid for other information.

Step 6 To display the pyramid, fold the flap under and secure with a paper clip, if needed.

Single-Pocket or One-Pocket Foldable® By Dinah Zike

Step 1 Using a large piece of paper on a vertical axis, fold the bottom edge of the paper upwards, about 5 cm.

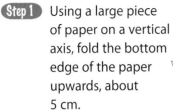

Step 2 Glue or staple the outer edges to make a large pocket.

PROJECT FORMAT
Use 11" × 17" or 12" × 18" paper and fold it vertically or horizontally to make a large pocket project.

Variations

A Make the one-pocket project using the paper on the horizontal axis.

B To store materials securely inside, fold the top of the paper almost to the center, leaving about 2–4 cm between the paper edges. Slip the Foldables through the opening and under the top and bottom pockets.

Multi-Tab Foldable® By Dinah Zike

Step 1 Fold a sheet of notebook paper in half like a *hot dog*.

Step 2 Open the paper and on one side cut every third line. This makes ten tabs on wide ruled notebook paper and twelve tabs on college ruled.

Step 3 Label the tabs on the front side and use the inside space for definitions or other information.

Variation

A Make a tab for a title by folding the paper so the holes remain uncovered. This allows the notebook Foldable to be stored in a three-hole binder.

Reference Handbook

PERIODIC TABLE OF THE ELEMENTS

Element — Hydrogen
Atomic number — 1
Symbol — H
Atomic mass — 1.01
State of matter

- Gas
- Liquid
- Solid
- Synthetic

A column in the periodic table is called a **group**.

A row in the periodic table is called a **period**.

The number in parentheses is the mass number of the longest lived isotope for that element.

Period	1	2	3	4	5	6	7	8	9
1	Hydrogen 1 H 1.01								
2	Lithium 3 Li 6.94	Beryllium 4 Be 9.01							
3	Sodium 11 Na 22.99	Magnesium 12 Mg 24.31							
4	Potassium 19 K 39.10	Calcium 20 Ca 40.08	Scandium 21 Sc 44.96	Titanium 22 Ti 47.87	Vanadium 23 V 50.94	Chromium 24 Cr 52.00	Manganese 25 Mn 54.94	Iron 26 Fe 55.85	Cobalt 27 Co 58.93
5	Rubidium 37 Rb 85.47	Strontium 38 Sr 87.62	Yttrium 39 Y 88.91	Zirconium 40 Zr 91.22	Niobium 41 Nb 92.91	Molybdenum 42 Mo 95.96	Technetium 43 Tc (98)	Ruthenium 44 Ru 101.07	Rhodium 45 Rh 102.91
6	Cesium 55 Cs 132.91	Barium 56 Ba 137.33	Lanthanum 57 La 138.91	Hafnium 72 Hf 178.49	Tantalum 73 Ta 180.95	Tungsten 74 W 183.84	Rhenium 75 Re 186.21	Osmium 76 Os 190.23	Iridium 77 Ir 192.22
7	Francium 87 Fr (223)	Radium 88 Ra (226)	Actinium 89 Ac (227)	Rutherfordium 104 Rf (267)	Dubnium 105 Db (268)	Seaborgium 106 Sg (271)	Bohrium 107 Bh (272)	Hassium 108 Hs (270)	Meitnerium 109 Mt (276)

Lanthanide series	Cerium 58 Ce 140.12	Praseodymium 59 Pr 140.91	Neodymium 60 Nd 144.24	Promethium 61 Pm (145)	Samarium 62 Sm 150.36	Europium 63 Eu 151.96
Actinide series	Thorium 90 Th 232.04	Protactinium 91 Pa 231.04	Uranium 92 U 238.03	Neptunium 93 Np (237)	Plutonium 94 Pu (244)	Americium 95 Am (243)

Legend:
- Metal
- Metalloid
- Nonmetal
- Recently discovered

	13	14	15	16	17	18
						Helium 2 He 4.00
	Boron 5 B 10.81	Carbon 6 C 12.01	Nitrogen 7 N 14.01	Oxygen 8 O 16.00	Fluorine 9 F 19.00	Neon 10 Ne 20.18
	Aluminum 13 Al 26.98	Silicon 14 Si 28.09	Phosphorus 15 P 30.97	Sulfur 16 S 32.07	Chlorine 17 Cl 35.45	Argon 18 Ar 39.95

10	11	12	13	14	15	16	17	18
Nickel 28 Ni 58.69	Copper 29 Cu 63.55	Zinc 30 Zn 65.38	Gallium 31 Ga 69.72	Germanium 32 Ge 72.64	Arsenic 33 As 74.92	Selenium 34 Se 78.96	Bromine 35 Br 79.90	Krypton 36 Kr 83.80
Palladium 46 Pd 106.42	Silver 47 Ag 107.87	Cadmium 48 Cd 112.41	Indium 49 In 114.82	Tin 50 Sn 118.71	Antimony 51 Sb 121.76	Tellurium 52 Te 127.60	Iodine 53 I 126.90	Xenon 54 Xe 131.29
Platinum 78 Pt 195.08	Gold 79 Au 196.97	Mercury 80 Hg 200.59	Thallium 81 Tl 204.38	Lead 82 Pb 207.20	Bismuth 83 Bi 208.98	Polonium 84 Po (209)	Astatine 85 At (210)	Radon 86 Rn (222)
Darmstadtium 110 Ds (281)	Roentgenium 111 Rg (280)	Copernicium 112 Cn (285)	* Ununtrium 113 Uut (284)	* Ununquadium 114 Uuq (289)	* Ununpentium 115 Uup (288)	* Ununhexium 116 Uuh (293)		* Ununoctium 118 Uuo (294)

* The names and symbols for elements 113-116 and 118 are temporary. Final names will be selected when the elements' discoveries are verified.

Gadolinium 64 Gd 157.25	Terbium 65 Tb 158.93	Dysprosium 66 Dy 162.50	Holmium 67 Ho 164.93	Erbium 68 Er 167.26	Thulium 69 Tm 168.93	Ytterbium 70 Yb 173.05	Lutetium 71 Lu 174.97
Curium 96 Cm (247)	Berkelium 97 Bk (247)	Californium 98 Cf (251)	Einsteinium 99 Es (252)	Fermium 100 Fm (257)	Mendelevium 101 Md (258)	Nobelium 102 No (259)	Lawrencium 103 Lr (262)

Glossary/Glosario

Multilingual eGlossary

A science multilingual glossary is available on the science Web site. The glossary includes the following languages.

Arabic
Bengali
Chinese
English
Haitian Creole
Hmong
Korean
Portuguese
Russian
Spanish
Tagalog
Urdu
Vietnamese

Cómo usar el glosario en español:
1. Busca el término en inglés que desees encontrar.
2. El término en español, junto con la definición, se encuentran en la columna de la derecha.

Pronunciation Key

Use the following key to help you sound out words in the glossary.

a	back (BAK)	ew	food (FEWD)
ay	day (DAY)	yoo	pure (PYOOR)
ah	father (FAH thur)	yew	few (FYEW)
ow	flower (FLOW ur)	uh	comma (CAH muh)
ar	car (CAR)	u (+ con)	rub (RUB)
e	less (LES)	sh	shelf (SHELF)
ee	leaf (LEEF)	ch	nature (NAY chur)
ih	trip (TRIHP)	g	gift (GIHFT)
i (i + com + e)	idea (i DEE uh)	j	gem (JEM)
oh	go (GOH)	ing	sing (SING)
aw	soft (SAWFT)	zh	vision (VIH zhun)
or	orbit (OR buht)	k	cake (KAYK)
oy	coin (COYN)	s	seed, cent (SEED, SENT)
oo	foot (FOOT)	z	zone, raise (ZOHN, RAYZ)

English — A — Español

absorption/amplitude

absorption: the transfer of energy from a wave to the medium through which it travels. (pp. 548, 584, 637)

acoustics: the study of how sound interacts with structures. (p. 585)

alternating current: an electric current that changes direction in a regular pattern. (p. 737)

amplitude modulation (AM): a change in the amplitude of a carrier wave. (p. 616)

amplitude: the maximum distance a wave varies from its rest position. (pp. 539, 573)

absorción/amplitud

absorción: transferencia de energía desde una onda hacia el medio a través del cual viaja. (páges. 548, 584, 637)

acústica: estudio de cómo interactúa el sonido con las estructuras. (pág. 585)

corriente alterna: corriente eléctrica que cambia la dirección en un patrón regular. (pág. 737)

amplitud modulada (AM): cambio en la amplitud de una onda portadora. (pág. 616)

amplitud: distancia máxima que varía una onda desde su posición de reposo. (páges. 539, 573)

broadcasting/echolocation (e koh loh KAY shun)   **radiodifusión/ecocolocación**

B

broadcasting: the use of electromagnetic waves to send information in all directions. (p. 615)

radiodifusión: uso de ondas electromagnéticas para enviar información en todas las direcciones. (pág. 615)

C

carrier wave: an electromagnetic wave that a radio or television station uses to carry its sound or image signals. (p. 616)

compression: region of a longitudinal wave where the particles of the medium are closest together. (pp. 532, 566)

concave lens: a lens that is thicker at the edges than in the middle. (p. 652)

concave mirror: a mirror that curves inward. (p. 645)

cone: one of many cells in the retina of the eye that respond to colors. (p. 657)

convex lens: a lens that is thicker in the middle than at the edges. (p. 652)

convex mirror: a mirror that curves outward. (p. 646)

crest: the highest point on a transverse wave. (p. 531)

onda portadora: onda electromagnética que una estación de radio o televisión usa para transportar señales de sonido o imagen. (pág. 616)

compresión: región de una onda longitudinal donde las partículas del medio están más cerca. (páges. 532, 566)

lente cóncavo: lente que es más grueso en los extremos que en el centro. (pág. 652)

espejo cóncavo: espejo que dobla hacia adentro. (pág. 645)

cono: una de muchas células en la retina del ojo que es sensible a los colores. (pág. 657)

lente convexo: lente que es más grueso en el centro que en los extremos. (pág. 652)

espejo convexo: espejo que curva hacia afuera. (pág. 646)

cresta: punto más alto en una onda transversal. (pág. 531)

D

diffraction: the change in direction of a wave when it travels by the edge of an object or through an opening. (p. 550)

diffuse reflection: reflection of light from a rough surface. (p. 644)

direct current: an electric current that continually flows in one direction. (p. 737)

Doppler effect: the change of pitch when a sound source is moving in relation to an observer. (p. 576)

difracción: cambio en la dirección de una onda cuando ésta viaja por el borde de un objeto o a través de una abertura. (pág. 550)

reflexión difusa: reflexión de la luz en una superficie rugosa. (pág. 644)

corriente directa: corriente eléctrica que fluye de manera continua en una dirección. (pág. 737)

efecto Doppler: cambio de tono cuando una fuente sonora se mueve con relación a un observador. (pág. 576)

E

echo: a reflected sound wave. (p. 584)

echolocation (e koh loh KAY shun): the process an animal uses to locate an object by means of reflected sounds. (p. 587)

eco: onda sonora reflejada. (pág. 584)

ecocolocación: proceso que un animal hace para ubicar un objeto por medio de sonidos reflejados. (pág. 587)

Glossary/Glosario • **G-3**

electric circuit/frequency

electric circuit: a closed, or complete, path in which an electric current flows. (p. 690)

electric conductor: a material through which electrons easily move. (p. 682)

electric current: the movement of electrically charged particles. (p. 690)

electric discharge: the process of an unbalanced electric charge becoming balanced. (p. 685)

electric generator: a device that uses a magnetic field to transform mechanical energy to electric energy. (p. 736)

electric insulator: a material through which electrons cannot easily move. (p. 682)

electric motor: a device that uses an electric current to produce motion. (p. 730)

electric resistance: a measure of how difficult it is for an electric current to flow in a material. (p. 692)

electromagnet: a magnet created by wrapping a current-carrying wire around a ferromagnetic core. (p. 728)

electromagnetic spectrum: the entire range of electromagnetic waves with different frequencies and wavelengths. (p. 609)

electromagnetic wave: a transverse wave that can travel through empty space and through matter. (pp. 535, 601)

circuito eléctrico/frecuencia

circuito eléctrico: trayectoria cerrada, o completa, por la que fluye corriente eléctrica. (pág. 690)

conductor eléctrico: material a través del cual se mueven los electrones con facilidad. (pág. 682)

corriente eléctrica: movimiento de partículas cargadas eléctricamente. (pág. 690)

descarga eléctrica: proceso por el cual una carga eléctrica no balanceada se vuelve balanceada. (pág. 685)

generador eléctrico: aparato que usa un campo magnético para transformar energía mecánica en energía eléctrica. (pág. 736)

aislante eléctrico: material por el cual los electrones no pueden fluir con facilidad. (pág. 682)

motor eléctrico: aparato que usa corriente eléctrica para producir movimiento. (pág. 730)

resistencia eléctrica: medida de qué tan difícil es para una corriente eléctrica fluir en un material. (pág. 692)

electroimán: imán fabricado al enrollar un alambre que transporta corriente alrededor de un núcleo ferromagnético. (pág. 728)

espectro electromagnético: rango completo de ondas electromagnéticas con frecuencias y longitudes de onda diferentes. (pág. 609)

onda electromagnética: onda transversal que puede viajar a través del espacio vacío y de la materia. (páges. 535, 601)

F

---

ferromagnetic (fer ro mag NEH tik) element: elements, including iron, nickel, and cobalt, that have an especially strong attraction to magnets. (p. 721)

focal length: the distance along the optical axis from the mirror to the focal point. (p. 645)

focal point: the point where light rays parallel to the optical axis converge after being reflected by a mirror or refracted by a lens. (p. 645)

frequency modulation (FM): a change in the frequency of a carrier wave. (p. 616)

frequency: the number of wavelengths that pass by a point each second. (pp. 542, 575)

elemento ferromagnético: elementos, incluidos hierro, níquel y cobalto, que tienen una atracción especialmente fuerte hacia los imanes. (pág. 721)

distancia focal: distancia a lo largo del eje óptico desde el espejo hasta el punto focal. (pág. 645)

punto focal: punto donde rayos de luz paralelos al eje óptico convergen después de ser reflejados por un espejo o refractados por un lente. (pág. 645)

frecuencia modulada (FM): cambio en la frecuencia de una onda portadora. (pág. 616)

frecuencia: número de longitudes de onda que pasan por un punto cada segundo. (páges. 542, 575)

gamma ray/longitudinal (lahn juh TEWD nul) wave　　　　　　　　**rayo gamma/onda longitudinal**

G

gamma ray: a high-energy electromagnetic wave with a shorter wavelength and higher frequency than all other types of electromagnetic waves. (p. 611)

Global Positioning System (GPS): a worldwide navigation system that uses satellite signals to determine a receiver's location. (p. 618)

grounding: providing a path for electric charges to flow safely into the ground. (p. 685)

rayo gamma: onda electromagnética de alta energía con longitud de onda más corta y frecuencia más alta que los demás tipos de ondas electromagnéticas. (pág. 611)

sistema de posicionamiento global (SPG): sistema mundial de navegación que usa señales satelitales para determinar la ubicación de un receptor. (pág. 618)

polo a tierra: suministrar una trayectoria para que las cargas eléctricas fluyan con seguridad hacia el suelo. (pág. 685)

H

hologram: a three-dimensional photograph of an object. (p. 665)

holograma: fotografía tridimensional de un objeto. (pág. 665)

I

infrared wave: an electromagnetic wave that has a wavelength shorter than a microwave but longer than visible light. (p. 610)

intensity: the amount of energy that passes through a square meter of space in one second. (p. 574)

interference: occurs when waves overlap and combine to form a new wave. (pp. 551, 576)

onda infrarroja: onda electromagnética que tiene una longitud de onda más corta que la de una microonda, pero más larga que la de la luz visible. (pág. 610)

intensidad: cantidad de energía que atraviesa un metro cuadrado de espacio en un segundo. (pág. 574)

interferencia: ocurre cuando las ondas coinciden y combinan para forma una onda nueva. (páges. 551, 576)

L

laser: an optical device that produces a narrow beam of coherent light. (p. 664)

law of reflection: law that states that when a wave is reflected from a surface, the angle of reflection is equal to the angle of incidence. (pp. 549, 643)

light: electromagnetic radiation that you can see. (p. 637)

longitudinal (lahn juh TEWD nul) wave: a wave in which the disturbance is parallel to the direction the wave travels. (pp. 532, 565)

láser: aparato óptico que produce un haz angosto y coherente de luz. (pág. 664)

ley de la reflexión: ley que establece que cuando una onda se refleja desde una superficie, el ángulo de reflexión es igual al ángulo de incidencia. (páges. 549, 643)

luz: radiación electromagnética que puede verse. (pág. 637)

onda longitudinal: onda en la que la perturbación es paralela a la dirección en que viaja la onda. (páges. 532, 565)

M

magnet: an object that attracts iron. (p. 717)
magnetic domain: region in a magnetic material in which the magnetic fields of the atoms all point in the same direction. (p. 722)
magnetic force: the force that a magnet applies to another magnet. (p. 718)
magnetic material: any material that a magnet attracts. (p. 721)
magnetic pole: the place on a magnet where the force it exerts is the strongest. (p. 718)
mechanical wave: a wave that can travel only through matter. (p. 531)
medium: a material in which a wave travels. (pp. 531, 566)
microscope: an optical device that forms magnified images of very small, close objects. (p. 662)
microwave: a low-frequency, low-energy electromagnetic wave that has a wavelength between about 1 mm and 30 cm. (p. 610)

imán: objeto que atrae al hierro. (pág. 717)
dominio magnético: región en un material magnético en el que los campos magnéticos de los átomos apuntan en la misma dirección. (pág. 722)
fuerza magnética: fuerza que un imán aplica a otro imán. (pág. 718)
material magnético: cualquier material que un imán atrae. (pág. 721)
polo magnético: lugar en un imán donde la fuerza que éste ejerce es la mayor. (pág. 718)
onda mecánica: onda que puede viajar sólo a través de la materia. (pág. 531)
medio: material en el cual viaja una onda. (páges. 531, 566)
microscopio: aparato óptico que forma imágenes aumentadas de objetos cercanos muy pequeños. (pág. 662)
microonda: onda electromagnética de baja frecuencia y baja energía que tiene una longitud de onda entre 1 mm y 30 cm. (pág. 610)

O

Ohm's law: a mathematical equation that describes the relationship between voltage, current, and resistance. (p. 694)
opaque: a material through which light does not pass. (p. 636)
optical device: any instrument or object used to produce or control light. (p. 661)

Ley de Ohm: ecuación matemática que describe la relación entre voltaje, corriente y resistencia. (pág. 694)
opaco: material por el que no pasa la luz. (pág. 636)
aparato óptico: cualquier instrumento usado para producir o controlar luz. (pág. 661)

P

parallel circuit: an electric circuit with more than one path, or branch, for an electric current to follow. (p. 702)
permanent magnet: a magnet that retains its magnetism even after being removed from a magnetic field. (p. 723)
photon: a particle of electromagnetic radiation. (p. 605)

circuito en paralelo: circuito eléctrico con más de una trayectoria, o rama, para que fluya una corriente eléctrica. (pág. 702)
imán permanente: imán que retiene su magnetismo incluso después de haberlo retirado de un campo magnético. (pág. 723)
fotón: partícula de radiación electromagnética. (pág. 605)

pitch/rod

pitch: the perception of how high or low a sound is; related to the frequency of a sound wave. (p. 565)

polarized: the condition of having electrons concentrated at one end of an object. (p. 683)

R
---

radiant energy: energy carried by an electromagnetic wave. (p. 601)

radio wave: a low-frequency, low-energy electromagnetic wave that has a wavelength longer than about 30 cm. (p. 610)

rarefaction (rayr uh FAK shun): region of a longitudinal wave where the particles of the medium are farthest apart. (pp. 532, 566)

reflecting telescope: a telescope that uses a mirror to gather and focus light from distant objects. (p. 662)

reflection: the bouncing of a wave off a surface. (pp. 548, 584, 637)

refracting telescope: a telescope that uses lenses to gather and focus light from distant objects. (p. 662)

refraction: the change in direction of a wave as it changes speed in moving from one medium to another. (pp. 550, 650)

regular reflection: reflection of light from a smooth, shiny surface. (p. 644)

resonance: an increase in amplitude that occurs when an object vibrating at its natural frequency absorbs energy from a nearby object vibrating at the same frequency. (p. 578)

reverberation: the collection of reflected sounds from the surfaces in a closed space. (p. 585)

rod: one of many cells in the retina of the eye that responds to low light. (p. 657)

tono/bastón

tono: percepción de qué tan alto o bajo es el sonido; relacionado con la frecuencia de la onda sonora. (pág. 565)

polarizado: condición de tener electrones concentrados en un extremo de un objeto. (pág. 683)

energía radiante: energía que transporta una onda electromagnética. (pág. 601)

onda de radio: onda electromagnética de baja frecuencia y baja energía que tiene una longitud de onda mayor de más o menos 30 cm. (pág. 610)

rarefacción: región de una onda longitudinal donde las partículas del medio están más alejadas. (páges. 532, 566)

telescopio de reflexión: telescopio que tiene un espejo para reunir y enfocar luz de objetos lejanos. (pág. 662)

reflexión: rebote de una onda desde una superficie. (páges. 548, 584, 637)

telescopio de refracción: telescopio que tiene lentes para reunir y enfocar luz de objetos lejanos. (pág. 662)

refracción: cambio en la dirección de una onda a medida que cambia de velocidad al moverse de un medio a otro. (páges. 550, 650)

reflexión especular: reflexión de la luz desde una superficie lisa y brillante. (pág. 644)

resonancia: aumento en la amplitud que ocurre cuando un objeto que vibra a su frecuencia natural absorbe energía de un objeto cercano y vibran a la misma frecuencia. (pág. 578)

reverberación: colección de sonidos reflejados de superficies en un espacio cerrado. (pág. 585)

bastón: una de muchas células en la retina del ojo sensible a la luminosidad baja. (pág. 657)

series circuit/ultraviolet wave circuito en serie/onda ultravioleta

S

series circuit: an electric circuit with only one closed path for an electric current to follow. (p. 701)

sonar: a system that uses the reflection of sound waves to find underwater objects. (p. 587)

sound wave: a longitudinal wave that can travel only through matter. (p. 565)

static charge: an unbalanced electric charge on an object. (p. 680)

circuito en serie: circuito eléctrico con sólo una trayectoria cerrada para que fluya una corriente eléctrica. (pág. 701)

sonar: sistema que usa la reflexión de ondas sonoras para encontrar objetos bajo el agua. (pág. 587)

onda sonora: onda longitudinal que sólo viaja a través de la materia. (pág. 565)

carga estática: carga eléctrica no balanceada en un objeto. (pág. 680)

T

temporary magnet: a magnet that quickly loses its magnetism after being removed from a magnetic field. (p. 723)

transformer: a device that changes the voltage of an alternating current. (p. 739)

translucent: a material that allows most of the light that strikes it to pass through, but through which objects appear blurry. (p. 636)

transmission: the passage of light through an object. (pp. 548, 637)

transparent: a material that allows almost all of the light striking it to pass through, and through which objects can be seen clearly. (p. 636)

transverse wave: a wave in which the disturbance is perpendicular to the direction the wave travels. (p. 531)

trough: the lowest point on a transverse wave. (p. 531)

turbine (TUR bine): a shaft with a set of blades that spins when a stream of pressurized fluid strikes the blades. (p. 738)

imán temporal: imán que rápidamente pierde su magnetismo después de haberlo retirado de un campo magnético. (pág. 723)

transformador: aparato que cambia el voltaje de una corriente alterna. (pág. 739)

translúcido: material que permite el paso de la mayor cantidad de luz que lo toca, pero a través del cual los objetos se ven borrosos. (pág. 636)

transmisión: paso de la luz a través de un objeto. (páges. 548, 637)

transparente: material que permite el paso de la mayor cantidad de luz que lo toca, y a través del cual los objetos pueden verse con nitidez. (pág. 636)

onda transversal: onda en la que la perturbación es perpendicular a la dirección en que viaja la onda. (pág. 531)

seno: punto más bajo en una onda transversal. (pág. 531)

turbina: eje con una serie de paletas que gira cuando un chorro de fluido a presión golpea las paletas. (pág. 738)

U

ultraviolet wave: an electromagnetic wave that has a slightly shorter wavelength and higher frequency than visible light. (p. 611)

onda ultravioleta: onda electromagnética que tiene una longitud de onda ligeramente menor y mayor frecuencia que la luz visible. (pág. 611)

vibration/X-ray　　　　　　　　　　　　　　　　　　　　　　　　　　　　　　**vibración/ rayo X**

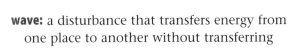

vibration: a rapid back-and-forth motion that can occur in solids, liquids, or gases. (p. 565)

voltage: the amount of energy used to move one coulomb of electrons through an electric circuit. (p. 693)

vibración: movimiento rápido de atrás hacia adelante que puede ocurrir en sólidos, líquidos o gases. (pág. 565)

voltaje: cantidad de energía usada para mover un culombio de electrones por un circuito eléctrico. (pág. 693)

wave: a disturbance that transfers energy from one place to another without transferring matter. (p. 529)

wavelength: the distance between one point on a wave and the nearest point just like it. (pp. 541, 575)

onda: perturbación que transfiere energía de un lugar a otro sin transferir materia. (pág. 529)

longitud de onda: distancia entre un punto de una onda y el punto más cercano similar al primero. (páges. 541, 575)

X

X-ray: a high-energy electromagnetic wave that has a slightly shorter wavelength and higher frequency than an ultraviolet wave. (p. 611)

rayo X: onda electromagnética de alta energía que tiene una longitud de onda ligeramente más corta y frecuencia más alta que una onda ultravioleta. (pág. 611)

Index

Absorption | *Italic numbers* = illustration/photo **Bold numbers** = vocabulary term *lab* = indicates entry is used in a lab on this page | Electric current

A

Absorption
  color and, 639, *639*
  explanation of, **548, 584,** 637
Academic Vocabulary, 552, 586, 618, 637, 699, 722. *See also* **Vocabulary**
Acoustics, 585, 585–586, *586*
Airport security, 621, *621*
Align, 722
Alternating current, 737
Ampere, 691, 694
Amplitude
  energy and, 539–540, *540*
  explanation of, **539,** 545, **573,** 574, 577, 583
  increase in, 578
  of longitudinal waves, *540, 1514*
  of transverse waves, 540, *540*
Amplitude modulation (AM), 616, 617
Analyze, 618
Angle of incidence, 643, 644, 648
Angle of reflection, 643, 644
Anvil, *568*
Atom(s)
  electrically charged, 680
  magnetic field of, 721, 722
  parts of, 679, *679,* 680
Aurora(s), 720, *720*

B

Battery(ies)
  explanation of, 699, *699*
  as source of electric energy, 700, *700*
  voltage of, 693, *693*
Beat(s), 577
Big Idea, 526, 556, 562, 592, 598, 626, 632, 670, 676, 708, 714, 744
  Review, 559, 595, 629, 673, 711, 747
Bird(s)
  navigation of, 721 *lab*

C

Camera(s), 663, *663,* 666
Careers in Science, 537
CD(s), 664, *664*
Cell phone(s), 617, *617*
Cell tower(s), 617, *617*
Chapter Review, 558–559, 594–595, 628–629, 672–673, 710–711, 746–747
Chemical energy
  transformed to electric energy, 700
Chemical reaction(s)
  in batteries, 699, 700, *700*

Circuit, 690
Circuit breaker(s), 703
Cloud(s)
  lightening and, 685, *686*
Cochlea, *568,* 569
Cochlear implant(s), 571, *571*
Color
  combination of, 639, *639*
  effects of, 641
  human eye and, 657, *657*
  light and, *638,* 638–639, 638 *lab,* 641
  wavelength and, 638, 639, *639,* 654, *654*
Color psychology, 641
Common Use. *See* **Science Use v. Common Use**
Commutator, 731, *731,* 737
Compact fluorescent lamp(s) (CFL), 700
Compass(es)
  explanation of, 720, *720,* 721 *lab,* 727
  to measure electric current, 733
Compression(s)
  amplitude and, 540
  explanation of, **532, 566,** *566,* 576
  wavelength and, 575
Computed tomography (CT) scanner(s), 622
Concave, 645
Concave lens(es)
  explanation of, **652,** 654, *654*
  vision correction and, 657
Concave mirror(s), 645, 645–646, *646*
Conduction
  transferring electric charge by, 684, *684,* 690
Conductor(s). *See* **Electric conductor(s)**
Cone cell, 657
Constructive interference, 552, 576
Contact lenses, 657
Converging lens(es), 652. *See also* **Convex lens(es)**
Convex lens(es)
  in cameras, 663, *663*
  cornea as, 656
  explanation of, 652, 653, *653*
  vision correction and, 657
Convex mirror(s), 646, *646*
Cornea, 656, *656,* 657
Coulomb, 691, 693
Crests, 531
Critical thinking, 536, 544, 553

D

Decibel (dB), 574
Decibel scale, 574, *574*
Density, 567

Destructive interference, 552, 576
Device, 699
Diffraction
  explanation of, **550**
  of sound waves and light waves, 551
Diffuse reflection, 644
Digital signal(s), 616
Direct current, 737
Distance
  electric force and, 681, *681*
Diverging lens(es), 652. *See also* **Concave lenses**
Dolphin(s), 587, *587*
Doppler effect, 575, 588
DVD(s), 664, *664*

E

Ear
  damage to human, 569, *569,* 571
  detection of sound by human, 568, *568*
Ear canal, *568*
Eardrum, *568,* 569
Earth
  magnetic field of, 720, *720*
  as source of electromagnetic waves, 605, *605,* 619
Earthquake(s)
  tsunamis and, 537
Echo(es), 584
Echolocation, 587
Electric charge(s)
  amount of, 681
  conduction and, 684, *684*
  contact and, 683, *683*
  explanation of, **679,** *679,* 680
  induction and, 683, *683*
  insulators and conductors and, 682, *682,* 692
  lightening and, 685, *686*
  positive and negative, 680, *680,* 683, 688
  Van de Graaff generator and, 688
Electric circuit(s)
  explanation of, **690**
  functions of, 691, *691,* 692 *lab,* 704 *lab*
  in home, 699 *lab,* 702–703, *703*
  parallel, 702, *702*
  parts of, 699, 699–700, *700*
  series, 701, *701*
  simple, 690
  voltage and, 693, *693,* 694, 695, 697
Electric conductor(s)
  electric resistance of, 692, 704
  explanation of, **682,** *682,* 683, *683*
Electric current
  in batteries, 693, *693*

I-2 • Index

Electric discharge
  danger of excess, 703
  direct and alternating, 737
  electric motors and, 706–707 lab
  explanation of, **690,** 691
  generation of, 735, *735*
  magnetic field around, *727,* 727–728, *728*
  measurement of, 733
  moving wire and, 736, *736*
  Ohm's law and, 694, *694,* 695, *695*
  safety and, 703, 704, 740
  unit for, 691
Electric discharge
  explanation of, **685**
  lightening and, 685, *685,* 686
Electric energy
  batteries as source of, 700, *700*
  in home, 702–703, *703,* 740
  mechanical energy converted to, 738, 742–743 lab
  transformation of, 690, 693, 699
  transformed to kinetic energy, 699
  voltage and, 694, 695
Electric field
  distance and, 681, *681*
  explanation of, 602, 603 lab, 680
Electric force(s)
  amount of charge and, 681, *681*
  distance and, 681, *681*
  explanation of, 680
Electric generator(s)
  electric current produced by, 737, *738*
  electric power plants and, 738, *738*
  explanation of, **736**
  simple, 736, *737*
Electric insulator(s)
  electric resistance of, 692
  explanation of, *682,* **682**
Electric meter, 702, *703*
Electric motor(s)
  explanation of, 706, **730**
  function of, 730–731
  simple, 730, *730*
  use of, 731
Electric resistance
  of conductors and insulators, 692
  explanation of, **692,** 692 lab, 704
  length and thickness of materials and, 692
  metal wires and, 700
  Ohm's law and, 694, *694,* 695, *695*
Electric shock, 704
Electricity
  magnetism and, 727 lab
Electromagnet(s)
  construction of, 728, *728*
  in electric motors, 730, *730*
  explanation of, **728,** 730 lab
  properties of, 729
  uses of, 729, *729*
Electromagnetic radiation, 635, 638
Electromagnetic spectrum, 609, *610–611*
Electromagnetic wave(s)
  absorption in, 548
  classification of, 609, *610–611*
  differences in, 609 lab
  energy of, 605, *605*
  explanation of, *535,* **535, 601**
  formation of, 602, *602*
  sources of, *604,* 604–605
  speed of, 603
  from Sun, 535
  types of, 535
  uses of, 609, 613, 615–622, *616–622,* 624–625 lab
  wavelength and frequency of, 603, 609, *610–611,* 616, *616*
Electron(s)
  electric charge and, 680
  electric current and, 691, *691*
  explanation of, 679, *679*
  negative charge of, 680, *680*
  transfer of, 682–685, *683,* 700
  Van de Graaff generator and, 688
Element(s)
  ferromagnetic, 721
Elevator
  design of, 706–707 lab
Energy
  amplitude and, 539–540, *540*
  of electromagnetic waves, 605, *605*
  explanation of, **529**
  radiant, 601
  sounds and, 539 lab
  of sound waves, 566, *573,* 573–574, 573 lab, *574*
  waves as source of, *529,* 529–530, *530*
Energy transfer
  by waves, 530, *530*
Energy transformation, 695
Eye(s)
  correcting vision in, 657
  detection of color by, 657
  explanation of, 656
  parts of, 656, *656*
Eyeglasses, 657

F

Fennec fox, 568, *568*
Ferromagnetic element(s)
  in electromagnets, 728, *728*
  explanation of, 721
Fluorescent lightbulb(s), 621, *621*
Focal length
  explanation of, **645**
  of lens, 653
Focal point
  explanation of, **645**
  in eye, 656
  of lens, 653
  of mirrors, 653
Foldables, 535, 539, 551, 557, 567, 578, 583, 593, 601, 609, 615, 627, 637, 644, 655, 666, 671, 681, 693, 703, 709, 718, 729, 736, 745
Force field(s), 602
Frequency modulation (FM), 616, 617

Frequency
  explanation of, *575,* **575,** 576, *576*
  fundamental, 577
  unit for, 542
  wavelength and, 541 lab, 542, *542*
  of waves, **542,** 545
Freshwater
  sound waves in, 567
Fundamental, 577, *577*
Fuses, 703, *703*

G

Gamma ray(s), 622, *622*
Gas(es)
  sound waves in, 566, 567
Generator(s)
  wind-power, 742–743 lab
Global positioning system (GPS), *618,* **618**
Grounded lightening rod(s), 685, *685*
Ground-fault circuit interrupter(s) (GFCI), 703
Grounding, 685

H

Hammer (ear), 568
Hearing
  explanation of, 568, *568*
  loss of, 569, *569*
Hertz (Hz), 542
High-voltage circuit(s), 739, 740
Hologram(s), *665,* **665**
How It Works, 607, 688
Hubble Space Telescope, 547, *547,* 661
Human(s)
  as source of electromagnetic waves, 604

I

Image(s)
  explanation of, **637**
  real, 646, 653
  virtual, 644, 646, 653
Incident ray(s), 643, 644, *644*
Index of refraction, 650, 651, *651*
Induction
  transferring electric charge by, 683, *683,* 690
Infrared wave(s)
  explanation of, *535,* **535,** 604, *604,* 618
  uses for, 619, *619*
Inner ear, 568, *568*
Insulator(s). *See* **Electric insulator(s)**
Intensity, 574
Interference
  constructive and destructive, 552
  explanation of, 551, *551,* **576**
  standing waves and, 552
Interpret Graphics, 536, 544, 553
Ion(s), 680
Iris, 656, *656*

K

Kepler, Johannes, 607
Key Concepts, 528, 538, 546
   Check, 529, 531, 532, 535, 542, 543, 548, 551, 552
   Summary, 556
   Understand, 536, 544, 553
Kinetic energy
   electric energy transformed to, 699

L

Lab, 554–555, 588–589, 624–625, 668–669, 706–707, 742–743. *See also* **Launch Lab; MiniLab; Skill Practice**
Laser(s)
   explanation of, *664,* **664,** 665
   uses for, *664,* 665, *665*
Launch Lab, 529, 539, 547, 565, 573, 583, 601, 609, 615, 643, 650, 661, 679, 690, 699, 717, 727, 735
Law of reflection
   demonstration of, 648
   explanation of, **549, 643,** 645
Lens(es)
   affect on light by, 659
   concave, 652, 654, *654*
   convex, 652, 653, *653,* 656, 663, *663,* 665
   explanation of, **652**
   in optical devices, 661, 662, *662*
   zoom, 663 *lab*
Lesson Review, 536, 544, 553, 570, 580, 589, 606, 612, 623, 640, 647, 658, 667, 687, 696, 705, 724, 732, 741
Light
   absorption of, 637, *637*
   color and, *638,* 638–639, 638 *lab,* *639,* 641
   detection of, 656, 656–657, *657*
   explanation of, **635**
   matter and, 636, *636*
   reflection of, 637, *637,* 643, 643 646, *644, 645, 646,* 650 *lab*
   refraction of, *650,* 650–651, *651*
   solar sails and, 607, *607*
   sources of, 635, *635*
   transmission of, 637, *637*
   travel method of, 636, *636*
   uses of, 619, *619*
   wavelengths of, 638
Light microscope, 662, *662,* 663
Light ray(s)
   explanation of, **643,** *643*
   in lenses, 652, 653, *653,* 654, *654*
   mirrors and, *645,* 646
   water and, 652 *lab*
Light wave(s). *See also* **Wave(s)**
   diffraction of, 551
   energy source of, 529, *529*
   explanation of, 604
   frequency of, 635
   reflection in, 549
   refraction of, 550
   transmission in, 548
Lightbulb(s), 621, *621*
   function of, 690 *lab,* 691, 700
Lightening
   causes of, 685, *686*
   energy released by, 679
   explanation of, 685
   safety tips during, 685
Liquid(s)
   sound waves in, 566, 567
Longitudinal wave(s)
   amplitude and energy of, 540, *540, 541,* 573
   explanation of, 532, *532,* **565**
   rarefaction in, 566, *566*
   wavelength of, 541, *541*
Loudness, 574
Loudspeaker(s), 729, *729*

M

Magnet(s)
   attraction of magnetic materials by, 723, *723*
   in compasses, 720, *720,* 727
   electric currents and, 728, *728*
   electric motors and, *730,* 730–731, *731*
   explanation of, **717,** 717 *lab,* 721, 722
   permanent, 723, *723*
   roller coasters and, 725
   temporary, 723, *723,* 729
   wires as, 727 *lab*
Magnetic brake(s), 725
Magnetic domain, *722,* **722**
Magnetic field(s)
   around currents, *727,* 727–728, *728*
   combining, *719,* 719
   of Earth, 720, *720*
   electric current produced by, 735, 735–738, *736, 737, 738*
   explanation of, 602, 603 *lab,* 606, 719, 721, 722, 727, 733, 735
Magnetic field line(s), 719
Magnetic force, *718,* **718**
Magnetic material(s)
   attraction of, 723, *723*
   explanation of, **721**
   magnetic domain in, 722, *722*
Magnetic pole(s)
   electric motors and, 730, 731, *731*
   explanation of, **718,** 720, *720*
   forces between, 718, *718*
   reversible, 729, 730, 731, *731*
Magnetism
   electricity and, 727 *lab*
Math Skills, 543, 559, 587, 589, 593, 595, 603, 629, 651, 658, 673, 695, 696, 711, 739, 741, 747
Matter
   how waves travel through, 531 *lab*
   interaction between waves and, *547,* 547–549, *548*
   light and, 636, *636*
Mechanical energy
   converted to electric energy, 738, 742–743 *lab*
   sources of, 738, 742

Mechanical wave(s)
   explanation of, **531,** 605
   longitudinal waves as, 532, *532*
   transverse waves as, 531, *531*
   types of, 534
   vibrations and, 533, *533*
Medical application(s)
   for ultraviolet light, 620, *620*
Medium, 531, 566, *566,* 567, 583, **653**
Metal(s)
   as electric conductors, 682
Microscope(s)
   light, 662, *662,* 666
Microwave(s)
   from communication satellites, 618, *618*
   explanation of, *617,* 617–618, *618*
Middle ear, 568, *568*
MiniLab, 531, 541, 549, 569, 577, 585, 603, 620, 638, 645, 652, 663, 682, 692, 721, 730, 737. *See also* **Lab**
Mirror(s)
   concave, *645,* 645–646, *646*
   convex, 646, *646*
   explanation of, 644
   plane, 644, *644,* 645 *lab*
   uses for, 643 *lab*
Modulate, 616
Mood
   color and, 641
Moon
   appearance of, 635
Multimeter, 694, *695*
Music
   beats in, 577
   resonance in, 578
   sound quality of, 578
Musical instrument(s)
   making your own, 590–591 *lab*
   types of, 579, *579,* 581

N

Neutron(s), 679, *679*
NGC 3603, 547
Night-vision goggles, 619, *619*
Noise pollution, 586, *586*
Normal, 549

O

Ørsted, Hans, 727 *lab*
Ohm (OHM), 692
Ohm, Georg, 694, 697
Ohm's law
   explanation of, **694,** 697, 701
   use of, 694, *695, 695*
Opaque object(s)
   color and, 638, *638*
   explanation of, *636,* **636**
Optical axis, 645
Optical device(s)
   explanation of, **661**
   types of, *661,* 662, *662*
Optical fiber(s), 619, 661 *lab,* 666, *666*

Optical illusion
design of, 668–37*lab*
Optical technology
cameras as, 663, *663*
lasers as, *664*, 664–665, *665*
for magnification, *661*, 661–662, *662*
optical fibers as, 666, *666*
Outer ear, 568, *568*
Overtone(s), 577, *577*

P

Parallel, 701
Parallel circuit(s), *702,* **702**
Pendulum(s), 735 *lab*
Percussion instrument(s), 579, *579*
Peripheral vision, 657
Permanent magnet(s), 723, *723*
PET scan, 622
Photon, 635
Pigment(s), 639, *639*
Pitch, 575, 576
Plane mirror(s)
explanation of, 644
image in, 645 *lab*
Polarized object(s), 683
Pollution
noise, 586, *586*
Positron emission tomography
Prevent, 586
Prism(s), 655, *655*
Proton(s)
explanation of, 679, *679*
positive charge of, 680, *680*
Pupil, 656, *656*

R

Radiant, 601
Radiant energy, 601
Radiation
electromagnetic, 635, 638
Radio transmission, 617, *617*
Radio wave(s)s
digital signals and, 616, *616*
explanation of, 615
function of, 616
transmission and reception of, 617, *617*
Rainbow(s), 655, *655*
Raindrop(s)
as source of energy for water waves, *528,* 529
Rarefaction(s)
amplitude and, 540
explanation of, **532, 566,** 576
wavelength and, 575
Ray(s), 643, *643*
Reading Check, 530, 533, 539, 540, 548, 549
Real image(s)
of lens, 653, *653*
of mirror, 646
Reflected ray(s)
explanation of, 643, 644, *644*
in mirrors, 645
Reflecting telescope, 662, *662*

Reflection
angle of, 643, 644
color and, 639
diffuse, 644
explanation of, **548,** 584, *584*
law of, 549, *549,* 643
of light, *637, 637, 643,* 643–646, *644, 645, 646*
regular, 644
total internal, 666
Refracting telescope, 662, *662*
Refraction
explanation of, *550,* **550, 650**
index of, *650,* 651, *651*
wavelength and, 654–655
Regular reflection, 644
Resonance, 578
Retina, *656,* 657
Reverberation, 585, 586
Review Vocabulary, 529, 588, 604, 635, 695, 721. *See also* **Vocabulary**
Roller coaster(s), 725

S

Satellite(s)
communication, 618, *618*
Science & Society, 641, 725
Science Methods, 555, 591, 625, 669, 707, 743
Science museum exhibit(s), 624–625 *lab*
Science Use v. Common Use, 549, 566, 618, 653, 690, 719. *See also* **Vocabulary**
Seawater
sound waves in, 567
Seismic wave(s), 534
Series circuit(s), *701,* **701**
Shadow(s), 636, *636*
Skill Practice, 545, 581, 613, 648, 659, 697, 733. *See also* **Lab**
Solar sail(s), 607, *607*
Solid(s)
sound waves in, 566, 567
Sonar, 587
Sound interference, *576,* 576–577
Sound
detection of, *568,* 568–569
direction of, 569 *lab*
energy and, 539 *lab*
frequency of, 575, *575*
intensity of, 574, *574*
quality of, 578
sources of, 565, *565,* 565 *lab*
speed of, 567, *567,* 585 *lab*
Sound wave(s). *See also* **Wave(s)**
absorption of energy of, 584
design of spaces to control, 586, *586*
diffraction of, 551
in ear, 568, *568*
echoes and, 584
energy of, 566, *573,* 573–574, *573 lab, 574*
explanation of, *534,* **565**
frequency and pitch of, 575, *575, 576, 576*

high-frequency, *587,* 587–588, *588*
reflection of, 549, 584, 585
reverberation and, 585
speed of, 543, *543*
transmission of, 566, *566,* 583
wavelength of, 575
Spacecraft(s)
speed of, 607, *607*
Spectrum
explanation of, **609**
Standardized Test Practice, 560–561, 596–597, 630–631, 674–675, 712–713, 748–749
Star(s)
as source of electromagnetic waves, 604, *604*
Static charge, 680
Step-down transformer(s), 739, *739*
Step-up transformer(s), 739, *739,* 740
Stirrup, 568
String instrument(s), 579
Study Guide, 556–557, 592–593, 626–627, 670–671, 708–709, 744–745
Sun
appearance of, 635
charged particles from, 720
electromagnetic waves from, 535
energy from, 601, 604, *604,* 605
ultraviolet waves from, 605, *605*
Supernova, 604, *604*

T

Telephone lines
function of, 661 *lab*
Telescope(s)
explanation of, 661, *661,* 666
reflecting, 662, *662*
refracting, 662, *662*
Television transmission, 617
Temperature
explanation of, **604**
speed of sound and, 567
Temporary magnet(s), 723, *723,* 729
Thermal camera(s), 619
Thermal energy
electric energy transformed to, 690, 694, 700, 701
fuses and, 703
from lightening, 679
produced by lightbulbs, 693
Thermal imaging, 619
Titov, Vasily, 537
Total internal reflection, 666
Transformer(s)
explanation of, 739, *739*
use of, 740, *740*
Translucent object(s)
color and, 639, *639*
explanation of, *636,* **636**
Transmission
color and, 639
explanation of, **548,** *637,* **637**
Transparent object(s)
color and, 639, *639*
explanation of, 636, *636*

Transverse wave(s)
index of refraction and, 651
light and, 636, 652
refraction and, 650, *650*, 651, *651*

Transverse wave(s)
amplitude of, 539, *540*
energy of, 540
explanation of, 531, *531*
wavelength of, 541, *541*

Trough(s), 531

Tsunami(s)
explanation of, 537
prediction of, 537, *537*

Turbine(s), *738,* **738**

U

Ultrasound
explanation of, 587, *587*
medical use of, 588, *588*

Ultrasound scanner(s), 588, *588*

Ultraviolet wave(s)
dangers associated with, 620
energy of, 605
explanation of, 535, 604
from flowers, 605, *605*
uses of, *620,* 620–621, *621*

V

Van de Graaff generator, 688, *688*

Vibration(s)
mechanical waves and, 533, *533*

Virtual image(s)
of lens, 653, *653*
of mirror, 644, 646

Vision
correction of, 657, *657*
human eye and, 656, *656*
peripheral, 657

Visual Check, 530, 534, 540, 548, 551

Vocabulary, 527, 528, 538, 546. *See also* **Academic Vocabulary; Review Vocabulary; Science Use v. Common Use; Word Origin**
Use, 536, 544, 553, 557

Vocal cords, 579

Voice
as instrument, 579, *579*

Volt(s), 694

Voltage
of batteries, 693, *693*
in circuits, 693
explanation of, **693,** 739
Ohm's law and, 694, *694,* 695, *695*

Voltmeter, 695

W

Water
as electric conductor, 703
light and, 652 *lab*

Water wave(s). *See also* **Wave(s)**
energy transfer by, 528, 529, 530, *530*
explanation of, *534*
reflection in, 549

Wave(s)
amplitude and, 539–540, *540,* 545
collision of, 547 *lab*
diffraction of, *550,* 550–551
electromechanical, 535, *535*
energy transfer by, 530, *530*
explanation of, **529**
frequency of, 542, *542,* 545
interaction between matter and, *547,* 547–549, *548*
interference of, *551,* 551–552, *552*
longitudinal, 532, *532*
mechanical, 531–534, *533, 534*
method to create, 529 *lab*
reflection of, 549, *549*

refraction of, 550, *550*
as source of energy, 529, *529*
speed of, 543, 554–555 *lab*
standing, 552, *552*
transverse, 531, *531*
tsunamis as, 537

Wave speed equation, 543

Wavelength
color and, 638, 639, *639,* 654, *654*
detection of light and, *656,* 656–657, *657*
of electromagnetic waves, 603, 609, *610–611,* 616, *616*
explanation of, **541,** *541,* 542, 545, *575, 575*
frequency and, 541 *lab,* 542, *542*
refraction and, 654–655

What do you think?, 527, 536, 544, 553

White light, 638, 639, 654

Wind instruments, 579, 581

Wind-power generator(s)
design of, 742–743 *lab*

Word Origin, 535, 539, 566, 578, 601, 609, 616, 685, 690, 701, 718, 731. *See also* **Vocabulary**

Writing In Science, 559, 595, 629, 673, 711, 747

X

X-ray(s)
for detection and security purposes, *621, 621*
medical uses for, 615 *lab,* 622, *622*

Z

Zoom len(s), 663, 663 lab

Credits

Photo Credits

COV Matt Meadows/Peter Arnold, Inc.; **ii** Matt Meadows/Peter Arnold, Inc.; **vii** Ransom Studios; **vii** Ransom Studios; **viii** altrendo images/Getty Images; **ix** Fancy Photography/Veer; **525** Warren Faidley/Corbis; **526–527** Kaz Mori/Getty Images; **528** Deco/Alamy; **529** Hutchings Photography/Digital Light Source; **531** Hutchings Photography/Digital Light Source; **534** (t)Royalty-Free/Corbis, (c)Jason Hosking/zefa/Corbis, (b)BluesandViews/Getty Images; **535** (l)Papilio/Alamy, (r)Ted Kinsman/Science Source; **536** Hutchings Photography/Digital Light Source; **537** (inset)NOAA, (bkgd)Comstock/PunchStock; **538** Patrick McFeeley/National Geographic Stock; **539** Hutchings Photography/Digital Light Source; **541** Hutchings Photography/Digital Light Source; **545** (1)McGraw-Hill Education, (3)Comstock Images/Alamy, (others)Hutchings Photography/Digital Light Source; **546** Gustoimages/Science Source; **547** (t)Hutchings Photography/Digital Light Source, (b)NASA, ESA, and the Hubble Heritage (STScI/AURA)-ESA/Hubble Collaboration; **548** Car Culture/Getty Images; **549** (t)McGraw-Hill Education, (b)Hutchings Photography/Digital Light Source; **550** (t)GIPhotoStock/Science Source, (bl)©sciencephotos/Alamy, (br)Educational Images LTD/Custom Medical Stock Photo/Newscom; **552** Andrew Lambert/Science Source; **553** Educational Images LTD/Custom Medical Stock Photo/Newscom; **554** (1)McGraw-Hill Education, (3)Comstock Images/Alamy, (others)Hutchings Photography/Digital Light Source; **555** Hutchings Photography/Digital Light Source; **556** GIPhotoStock/Science Source; **559** Kaz Mori/Getty Images; **562–563** Brad Kaye/Getty Images; **564** Mike Kemp/Getty Images; **565** Hutchings Photography/Digital Light Source; **567** JG1153/Getty Images; **568** Michael Mährlein/age fotostock; **569** (t) Hutchings Photography/Digital Light Source, (b)Wave Royalty Free/age fotostock; **570** Michael Mährlein/age fotostock; **571** (t)Elizabeth Hoffmann/Getty Images, (b)Steve Hamblin/Alamy; **572** Stefano Cellai/age fotostock; **573** Hutchings Photography/Digital Light Source; **577** Hutchings Photography/Digital Light Source; **581** (4)McGraw-Hill Education, (others) Hutchings Photography/Digital Light Source; **582** (inset)Emory Kristof/National Geographic Stock, (bkgd)NOAA; **583** Hutchings Photography/Digital Light Source; **585** Hutchings Photography/Digital Light Source; **586** (t)Purestock/SuperStock, (bl)View Pictures Ltd/SuperStock, (br)Zave Smith/Getty Images; **587** DAJ/Getty Images; **588** Zephyr/Science Source; **589** Purestock/SuperStock; **590** (2 3)McGraw-Hill Education, (others)Hutchings Photography/Digital Light Source; **591** Hutchings Photography/Digital Light Source; **592** (t)Michael Mährlein/age fotostock, (b)Zave Smith/Getty Images; **595** Brad Kaye/Getty Images; **598–599** NASA; **600** Menno Boermans/Aurora Photos/Corbis; **601** Hutchings Photography/Digital Light Source; **603** Hutchings Photography/Digital Light Source; **604** X-ray: NASA/CXC/J.Hester (ASU); Optical: NASA/ESA/J.Hester & A.Loll (ASU); Infrared: NASA/JPL-Caltech/R.Gehrz (Univ. Minn.); **605** (t c)Bjorn Rorslett/Science Source, (bl)Hutchings Photography/Digital Light Source, (br)Ralf Nau/Getty Images; **606** X-ray: NASA/CXC/J.Hester (ASU); Optical: NASA/ESA/J.Hester & A.Loll (ASU); Infrared: NASA/JPL-Caltech/R.Gehrz (Univ. Minn.); **607** NASA; **608** Felix Stenson/age fotostock; **609** Hutchings Photography/Digital Light Source; **610** (t)Yury Minaev/Getty Images, (c)PM Images/Getty Images, (b)D. Hurst/Alamy; **611** (l)Digital Vision/PunchStock, (tr)Peter Cade/Getty Images, (br)©Horizon International Images Limited/Alamy; **612** (t)Felix Stenson/age fotostock, (c)BananaStock/PunchStock, (b)©Horizon International Images Limited/Alamy; **613** (tl)Steve Cole/Getty Images, (cl) McGraw-Hill Education, (bl r)Hutchings Photography/Digital Light Source; **614** Don Farrall/Getty Images; **615** Thinkstock Images/Getty Images; **617** Joe Drivas/Getty Images; **619** (t)Simon Belcher/Alamy, (c)NASA, (b)Corbis/age fotostock; **620** (t)Daniel L. Geiger/SNAP/Alamy, (b)Digital Vision/age fotostock; **621** Ken Cavanagh/McGraw-Hill Education; **623** NASA; **624** (l) Nikreates/Alamy, (tr)NASA/JPL, (br)Getty Images; **625** Hutchings Photography/Digital Light Source; **626** (t)D. Hurst/Alamy, (b)Ken Cavanagh/McGraw-Hill Education; **629** NASA; **632–633** Charles Krebs/Getty Images; **634** McGraw-Hill Education; **635** (t)Hutchings Photography/Digital Light Source, (b)Andrew Geiger/Getty Images; **636** (t)Hutchings Photography/Digital Light Source, (b)Jasper James/Getty Images; **637** Picturenet/Getty Images; **640** (t)Jasper James/Getty Images, (b)Picturenet/Getty Images; **641** (t)Tanya Constantine/Getty Images, (c)image100/Corbis, (b)Simon Stuart-Miller/Alamy; **642** Jeremy Horner/Getty Images; **643** Hutchings Photography/Digital Light Source; **645** Hutchings Photography/Digital Light Source; **646** sciencephotos/Alamy; **648** Hutchings Photography/Digital Light Source; **649** Image Source/Getty Images; **650** Hutchings Photography/Digital Light Source; **652** (tl tr)Don Farrall/Getty Images, (b) Hutchings Photography/Digital Light Source; **655** Getty Images; **656** darren baker/Getty Images; **657** MIXA/Getty Images; **658** Getty Images; **659** (1 3 4)McGraw-Hill Education, (2 5)Hutchings Photography/Digital Light Source; **660** Michael Kappeler/DDP/Getty Images; **661** (t)Hutchings Photography/Digital Light Source, (b)StockTrek/Getty Images; **663** Hutchings Photography/Digital Light Source; **664** Peter Macdiarmid/Staff/Getty Images News/Getty Images; **665** (tl)AndreyPopov/Getty Images, (tr)James Leynse/Corbis, (bl)Jorge Fajl/Getty Images, (br)Construction Photography/Corbis; **666** Bruce Ando/Getty Images; **667** (t)AndreyPopov/Getty Images, (b)Bruce Ando/Getty Images; **668** (1 3 5)McGraw-Hill Education, (2 4 7) Hutchings Photography/Digital Light Source, (6)Steve Cole/Getty Images; **669** Hutchings Photography/Digital Light Source; **670** (t to b)Hutchings Photography/Digital Light Source, sciencephotos/Alamy, Don Farrall/Getty Images, Bruce Ando/Getty Images; **673** Charles Krebs/Getty Images; **676–677** Bettmann/Corbis; **678** Jurkos/Getty Images; **679** Hutchings Photography/Digital Light Source; **682** Hutchings Photography/Digital Light Source; **683** Hutchings Photography/Digital Light Source; **685** (l)vladimir zakharov/Getty Images, (inset)Okan Metin/iStock/360/Getty Images; **686** (b)Alexey Stiop/Alamy, (bkgd)Ralph H Wetmore II/Getty Images; **687** Jurkos/Getty Images; **689** Construction Photography/Corbis; **690** Hutchings Photography/Digital Light Source; **691** Hutchings Photography/Digital Light Source; **692** Hutchings Photography/Digital Light Source; **696** Construction Photography/Corbis; **697** (t bll)McGraw-Hill Education, (cl r)Hutchings Photography; **698** RichardBakerUSA/Alamy; **699** Hutchings Photography/Digital Light Source; **700** Hutchings Photography/Digital Light Source; **703** Tetra Images/Getty Images; **704** Hutchings Photography/Digital Light Source; **705** Hutchings Photography/Digital Light Source; **706** (1 2 5 8) Hutchings Photography/Digital Light Source, (others)McGraw-Hill Education; **707** Hutchings Photography/Digital Light Source; **708** RichardBakerUSA/Alamy; **711** Bettmann/Corbis; **714–715** age fotostock/SuperStock; **716** imagebroker/Alamy; **717** (tl)Joel Arem/Science Source, (tr)

Credits

Clive Streeter/Getty Images, (b)Stockbyte/Getty Images; **718** Hutchings Photography/Digital Light Source; **719** Cordelia Molloy/Science Source; **720** Grambo/Getty Images; **721** Hutchings Photography/Digital Light Source; **724** (t)Cordelia Molloy/Science Source, (b)Joel Arem/Science Source; **726** Alexis Rosenfeld/Science Source; **727** Hutchings Photography/Digital Light Source; **730** Hutchings Photography/Digital Light Source; **733** (1 3) McGraw-Hill Education, (others)Hutchings Photography/Digital Light Source; **734** (inset)bobo/Alamy, (bkgd)BlueMoon Stock/Alamy; **735** Hutchings Photography/Digital Light Source; **738** (t)Harris Shiffman/iStock/Getty Images, (b)Monty Rakusen/Cultura/Getty Images; **741** (t)Harris Shiffman/iStock/Getty Images, (b)age fotostock/SuperStock; **742** (2 4 5) McGraw-Hill Education, (others)Hutchings Photography/Digital Light Source; **743** Hutchings Photography/Digital Light Source; **744** (t) imagebroker/Alamy, (c)Alexis Rosenfeld/Science Source, (b)bobo/Alamy; **746** Cordelia Molloy/Science Source; **747** (l)Cordelia Molloy/Science Source, (r)age fotostock/SuperStock; **SR0–SR1** Gallo Images - Neil Overy/Getty Images; **SR2** Hutchings Photography/Digital Light Source; **SR6** Michelle D. Bridwell/PhotoEdit; **SR7** (t)McGraw-Hill Education, (b)Dominic Oldershaw; **SR8** StudiOhio; **SR9** Timothy Fuller; **SR10** Aaron Haupt; **SR12** KS Studios; **SR13** Matt Meadows.